Introduction to QGIS
Open Source Geographic Information System

Scott Madry, Ph.D.

Tutorial Series

Credits & Copyright

Introduction to QGIS

Open Source Geographic Information System

by Scott Madry, Ph.D.

Published by Locate Press LLC

COPYRIGHT © 2021 LOCATE PRESS LLC
ISBN: 978-1734464306

Direct permission requests to `info@locatepress.com` or mail:
Locate Press LLC, PO Box 671897, Chugiak, AK, USA, 99567-1897

Publisher Website `http://locatepress.com`
Book Website `http://locatepress.com/itg`

No part of this work may be reproduced or transmitted in any form or by any means, electronic or mechanical, including photocopying, recording, or by any information storage or retrieval system, without the prior written permission of the copyright owner and the publisher.

Contents

Foreword

Dear Reader

The QGIS project was started by Gary Sherman in 2002. What started back then as a simple selfless act of sharing his work has, in the intervening years, become an indispensable foundation for the daily workflows of many people around the world. They use it for their spatial data management, analysis and reporting needs. In tandem with the development of the actual QGIS software, there has been an upwelling of excellent community driven supporting services and materials around the QGIS project. This book is a fine example of what members of our user community have contributed to the ecosystem. Scott Madry, the author of this book, is a leading figure in the geospatial industry, with many years of deep experience to share. Scott typifies the ethos that makes a project like QGIS work. He has a deep passion for geospatial technologies and their role as a tool for sustainable development, social upliftment and human resilience, and he shares his time and knowledge generously. In this book Scott provides a robust and accessible framework for all who wish to get to grips with using QGIS as an on-ramp into the world of geospatial data management and analysis.

As you work your way through this book take a moment to reflect that if you take on board the knowledge that Scott has to share, you will be in a position to not only use QGIS like a pro, but become part of the community that exists around the QGIS.org project. The project exists and thrives based on the simple premise that we help each other by sharing our knowledge and time. Perhaps you can spend a few moments helping someone in your community to get started solving their problems with QGIS. Perhaps you are able to translate the QGIS interface into your own language, or report a bug when you encounter an issue. Whatever your contribution we welcome it and we welcome you into the community of users. With this book as your trusty companion you are in the best possible hands to make your way into the wonderful world of geospatial!

Regards

Tim Sutton

Tim Sutton

Author's Introduction

This is a hands-on tutorial using the Free and Open Source QGIS Geographic Information System software. I have been involved in Open Source GIS and Remote Sensing for many years, going back to the early days of GRASS GIS[1]. When I was working at the NASA Stennis Space Center's Institute for Technology Development's Space Remote Sensing Center in the mid-1980's we hosted the third GRASS user group meeting, and distributed GRASS on CD-ROMS, and also distributed the GRASS global GIS dataset. I was on the GRASS steering committee, and also was one of the founding members of the board of what has become the OGC, the Open Geospatial Consortium, and continued my research and teaching at Rutgers University, the International Space University, and now at the University of North Carolina at Chapel Hill. So my involvement in this community goes back a long way, and I have seen many improvements and advances in the tools, data, and our ability to use these for practical and useful purposes. It has been an interesting ride.

I have given many GIS university classes and short courses over the years around the world, well over two hundred on six of the seven continents, and I know how hard it is to gain and maintain a working knowledge of complex GIS software from a few days of short course instruction. It is just really difficult. I have seen many highly motivated students take my short course, go back to their "real world" and in a few months, if they do not quickly exercise their new skills, lose their ability to do useful work.

The goal of this guide is to create a document that is detailed enough that you can use it both for the initial learning process, but can also refer back to it, if you have let your skills atrophy and need to get back into using the tools, or need a refresher on a specific task like georeferencing or printing maps. I have used this in many of my classes and short courses, and have found that it is a good mix of detail and general concepts for new QGIS users. Please remember that there is much, much more to QGIS than what is presented here, this is simply the entry into a much larger domain. I hope it helps you! Feel free to contact me if you have questions, ideas, or suggestions; or if you find any errors. The key thing is just DO IT!

-Scott Madry, Ph.D. Chapel Hill, NC, USA 2021

madrys@email.unc.edu `http://scottmadry.web.unc.edu`

Note: I use a Mac, so your icons and windows may look slightly different if you use another operating system.

[1] `https://grass.osgeo.org/`

1. What is QGIS?

QGIS is an Open Source Geographic Information System (GIS) first developed by Gary Sherman of Alaska in 2002. The basic goal was to create an open source data viewer/GIS that was free and simple to learn and use, without the cost and difficulty of existing commercial GIS systems. It has grown significantly, and now runs on most Unix platforms, Windows, MacOS, and the Android operating systems.

QGIS is written in C++, using the Qt library (pronounced "cute"), and is tightly integrated with Python. All this makes it significantly smaller in size while using less RAM, and thus it is faster and easier to run on laptops than existing commercial GIS software. QGIS can even be run, with a complete database, directly from a flash drive. It can incorporate hundreds of plugins and actually uses many other Open Source capabilities, giving it both flexibility and power.

QGIS is released under the GNU Public License (GPL) Version 2 (`http://www.gnu.org/`), allowing users to view, modify, and share the source code.

1.1 The Origins of QGIS

QGIS started out as a simple data viewer, but it has grown significantly, and can now meet the GIS needs of many users. Its popularity was originally due, at least in part, to the fact that it used the ESRI shapefile as the vector format, GeoTIFF for rasters, could be extended with Python plugins.

`https://www.esri.com/library/whitepapers/pdfs/shapefile.pdf`

1.2 Data Formats

In QGIS 3.x and above, the standard format is the GeoPackage (.gpkg), which is a self-contained SQLite database that can house vectors, and rasters in a single file.

`https://www.GeoPackage.org`

QGIS continues to support shapefiles, along with a host of raster and vector formats through the use of the GDAL/OGR library.

`https://www.osgeo.org/projects/gdal/`

An issue with the shapefile format is that it requires a minimum of three files to create one GIS layer. If one of these is moved or lost, you can't load the layer. GeoPackage data are contained in a single file. Because of this, it is very good for use both data management and distribution.

QGIS 3.x was a major upgrade, and now uses the current Qt, Python, and other libraries, and it incorporates many upgrades, improvements, and bug fixes. Qt is a free and open source widget toolkit for creating graphical user interfaces as well as cross-platform applications that run on a number of software and hardware platforms such as Linux, Windows, MacOS, Android or embedded systems with little or no change in the underlying source code. It is the origin of the Q in QGIS, which was originally called Quantum GIS, but now it is simply QGIS. Some say "Q-jis"", some say "Q-G-I-S".

1.3 OSGeo—The Open Source Geospatial Foundation

QGIS is an official project of the Open Source Geospatial Foundation (`https://www.osgeo.org`). OSGeo is a non-profit foundation that exists to provide various services, including financial, organizational, and legal support to the geospatial open source community. There are several other projects and initiatives under the OSGeo banner, including GRASS GIS.

The basic idea of open source is that anyone can play, and access is not determined by money, location, or language. It is all about open standards, open data, open education, and open science. Open source fosters transparency, in that we can see the code we are using, and alter and improve it. It is also about the free transfer of knowledge and capabilities. GIS is no longer limited to only those who can afford expensive commercial licenses.

1.4 Translation into Other Languages

Open source software isn't limited to those who are fluent in English. International users are able to use QGIS as it has been, or is being, translated into some eighty-seven languages thus far. The chart below shows all translations that are under way at this time, and the percent completed:

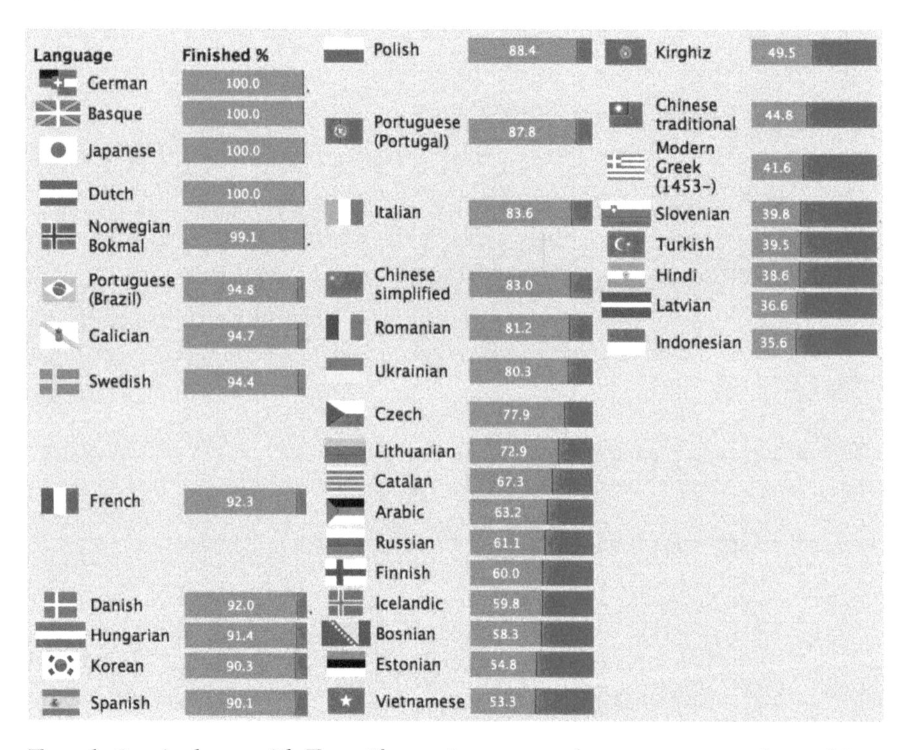

`https://www.qgis.org/en/site/getinvolved/translate.html`

Translation is done with Transifex, using a crowd-source approach, and is open to anyone who would like to participate. If you are fluent in a language other than English, please consider helping in this important effort. Open source also means that we are in charge. The community of developers and users are in charge of development directions, and we can build capabilities for user communities that are not commercially attractive, such as niche academic users like archaeologists or field biologists. Students can learn and use the open source tools in their own languages, and then take these tools with them when they leave school or start new ventures.

In order to change the language interface, click on `Settings->Options->General` in the menu. This is where you can set all your general user preferences, but it is probably best for beginning users not to change anything. Click on *Override system locale* option at the very top, and you will be able to choose the language interface you want to use. Be aware that some are not complete, so you might see a mixture of the selected language and English. The documentation now is all in English, but there are several tutorials in different languages.

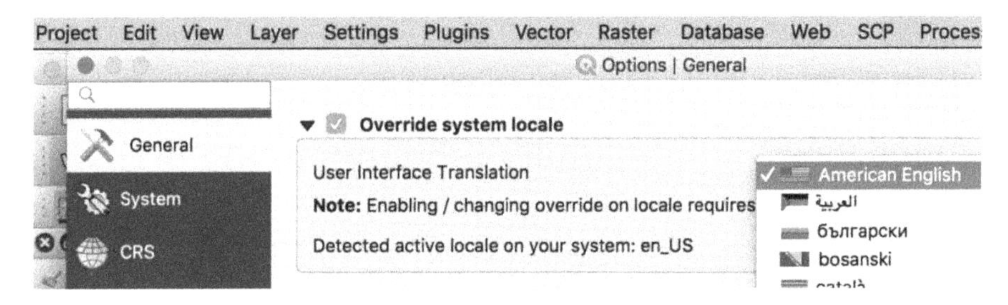

QGIS is also very much also a community, and part of a larger community. It is a community of developers and users, who work collaboratively around the world. The developers work collaboratively in planning future developments and bug fixes. There are many user communities as well, ranging from archaeology to zoology and everything in between. Open source also means that we can actually see the code, and know what the tools are doing, and that we can share not only ideas and data, but can also share and improve the capabilities. QGIS is free. Free as in lunch (no cost) and also free as in liberty (free to use, share, and alter).

1.5 What you get when you Download QGIS

When you download QGIS, you get the full desktop version, which includes GRASS GIS and several other Open Source tools. There used to be a slimmed down browser version, but this is now integrated into the application. You can also download separately QGIS Server, which allows you to publish your QGIS projects and data on the web as a WMS and WFS services. It is a standards-compliant WMS 1.3 server. Finally, there is the QGIS Web Client, which allows you to access QGIS projects and data over the web. It is based on OpenLayers and GeoExt.

1.6 Download

Please go to the QGIS download site and download the current version for your computer: `https://www.qgis.org/en/site/forusers/download.html` QGIS has a dual-track approach to versions. Developer versions are released rapidly, as new capabilities and bug fixes are ready. There is also a long-term release (LTR) version which is supported and available for up to at least a year. QGIS is available on Windows, MacOS X, Linux, BSD, and Android platforms, including tablets. This tutorial uses QGIS LTR version 3.16. Latest versions are named after cities and towns around the world, and 3.18 is called Zurich.

If you use Windows and MacOS, the download is very simple. For Linux and BSD users, follow the directions provided on the website. Please delete any previous versions of QGIS before you do the installation. There are several training datasets available from the download website. Feel free to download them, but they are not required for this tutorial.

1.7 Data for the Hands-on Exercises

The datasets for this tutorial, and the tutorial documents, PowerPoint presentations, and other materials are available at:

```
https://locatepress.com/itq/QGIS_Intro_Data.zip
```

The files are frequently updated, so check back for new and updated content. Lets get started.

2. The QGIS User Interface

2.1 Getting Started

We will begin learning QGIS by using a GIS database from the Commando Drift Nature Reserve (Kommandodrif-natuurreservaat) in the Eastern Cape Province, South Africa. It is a wildlife park of about 6,000 hectares. The dam and lake were built in 1956, and the dam is 518 meters long. The database covers an area 16 KM E-W by 10 Km N-S. The database was originally in latitude/longitude and used the WGS84 geodetic datum, and is centered on 32° 04'S and 26° 02'E, with an elevation of about 1,000 meters above sea level. Our database has been converted to a metric UTM coordinate system, in UTM Zone 35 South.

https://www.sa-venues.com/game-reserves/commando-drift.php

See https://docs.qgis.org/3.16/en/docs/gentle_gis_introduction/ for a good introduction to coordinate systems in GIS (for example: WGS84, UTM).

The very first thing to do is open a new word processing document or text file and save it in your GIS workspace. Name it something like QGIS_worklog.doc. The idea is to have this open throughout your GIS work, or when you are working with your GIS projects to keep a running log of what you do, ideas, questions, problems, etc. You will do a GIS activity, switch to your log document, and briefly type what you did. It seems a bother, and it slows you down, but you should always do this when working on a project, as it is impossible to remember what you did a month or year later—what you named new files, where you put them, etc. It is a good habit to develop, so please do this. You'll save yourself many problems down the road. Also, be sure to save your project frequently as you work on it, as if it hangs or if you have other problems, it will revert to where you last saved your project, not where you are. There is no auto-save function in QGIS, so be sure to save your project after each new file is created. Document your work and save the data remotely, and always have a recent copy of your data backed up.

2.2 The QGIS Documentation Website

Next, open a new browser window and go to: http://www.qgis.org/en/docs/ This is the documentation page for QGIS, with the user guide, training manual, etc. At the time of this writing, the documentation is for version 3.16 and 3.10. Click on the *QGIS User Guide* and you can refer to it when you have questions about specific modules that are not covered here. Note that this is also the site for the *QGIS Training Manual*, which is an excellent source for additional training on all the QGIS functions once you complete this tutorial. It is an excellent idea to have two monitors when you are doing GIS. Use one for your QGIS display, and the other for your work log, pop-up windows, web browser for the users manual, etc. You can do it with one monitor, but I have found two monitors to be very helpful. I put mine one above the other, some people like them side by side. It is up to you, but try using two monitors.

Also be aware that QGIS is a very fast developing environment, with new capabilities added frequently. The QGIS project publishes a change log for each new version, which lists all the changes and updates. The change log for 3.16 can be found here: `https://qgis.org/en/site/forusers/visualchangelog316/index.html`

Start QGIS on your machine. You will soon see the QGIS interface, including a 'helpful hint' that is always shown (you can check a box to turn off these hints if they get old). It also shows you the most recent QGIS projects you have used. To open one of these, simply click on it. If the path to the data is not available, the project will be grayed out. The first time you start QGIS, you'll be presented with a blank page.

2.3 An Introduction to the QGIS User Interface

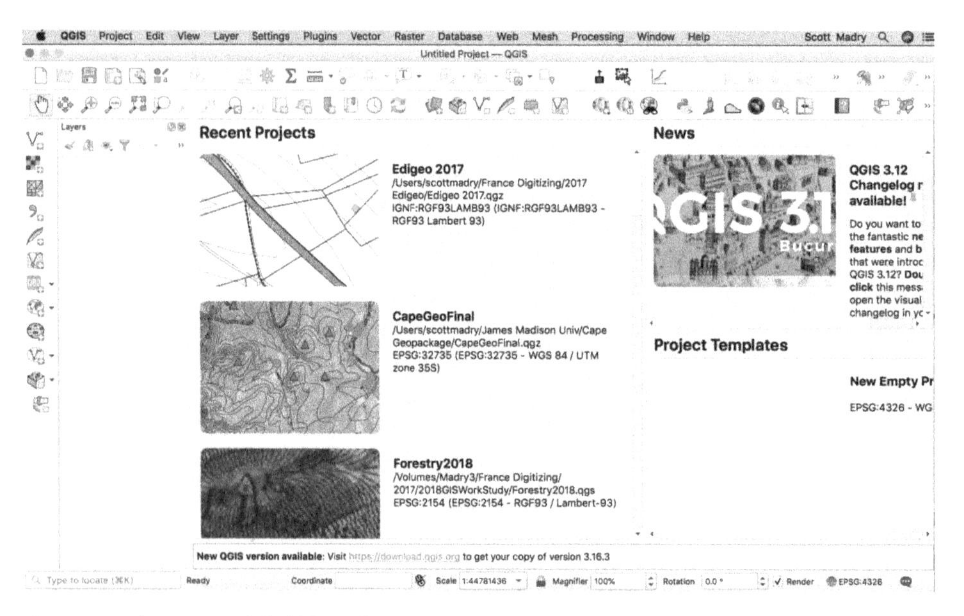

Once you have used QGIS and saved one or more projects, they will show in the *Recent Projects* panel as shown above. You can click on one and it will open it. To open a new project, as we will do now, go to the `Project` menu, and click on `New`.

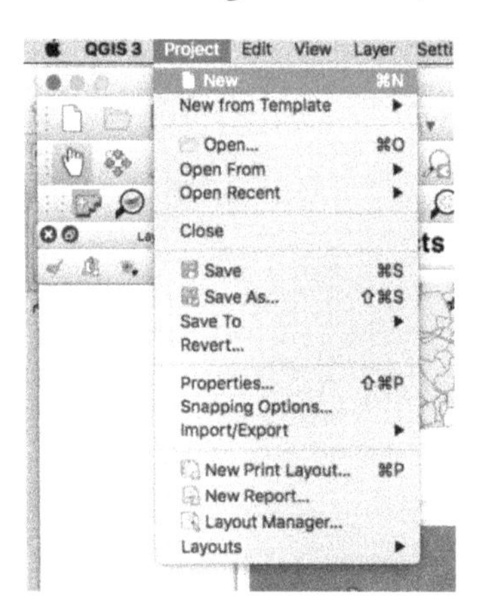

You can also click on `Project->Open Recent` to see all of the recently opened projects, and can click on any to open it. You will see the QGIS user interface, with an empty white space at right, which is where the data will be displayed.

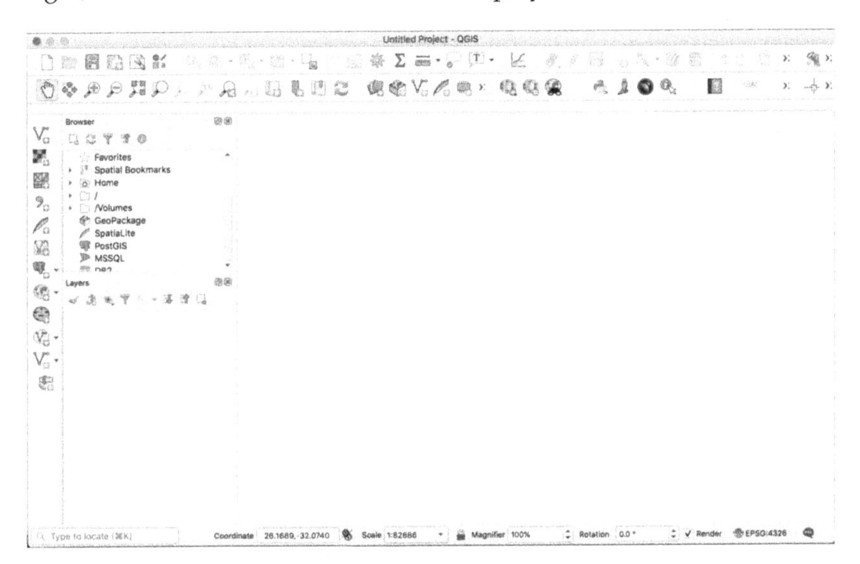

At left is the *Layers* panel, similar to the ArcGIS Table of Contents (TOC), where all open layers are displayed as a list. You can click each layer off and on by checking the boxes at left (once you add data). Below this are two tabs, for your *Browser Panel*, where you can search your computer folders for GIS files, and also for your layers, once you put data into your new project. This is shown below:

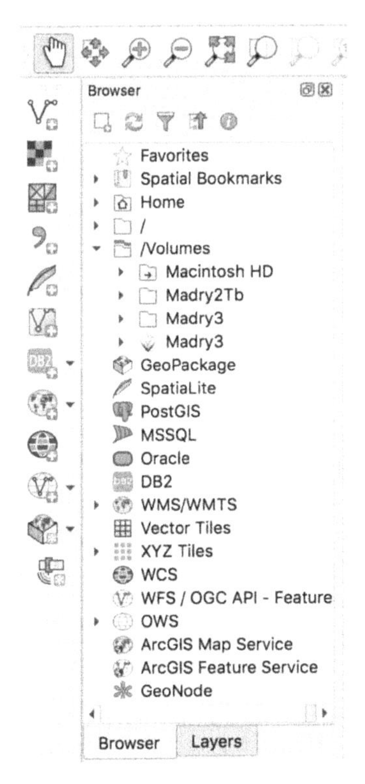

You can see that you can browse for data on your computer. If you ever accidentally delete the *Layers* panel or *Browser* panel, don't worry. Click on `View->Panels` menu and you will see all the various panel options that you can add or remove. Click on `Layers` and it will reappear.

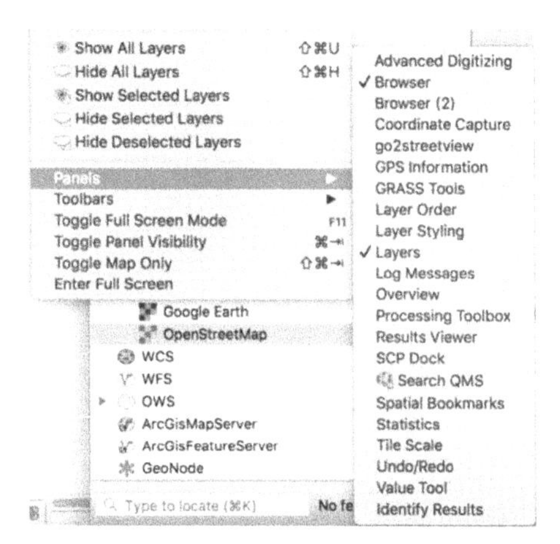

It may appear as a floating panel, but you can dock (attach) it by moving it to the left, top, bottom, or right part of your screen. You can see here that I have my *Browser* and *Layer* panels shown, and that there are many others. Most will pop up and appear at the appropriate time when you do certain actions, but you can always open any of them at any time. Remember that you can grab the border of these with the map display to move the border left or right, as you wish.

2.4 The QGIS Browser Panel

The QGIS *Browser* panel lets you navigate in your computer's file system and lets you manage the loading of geodata files within QGIS. You can access various formats of files (shapefiles, GeoDatabases, etc.), manage web-based WMS/WFS connections, and view various GRASS and other files. You can enter data into your workspace by simply dragging files from the *Browser* panel into your display panel (map area) at right.

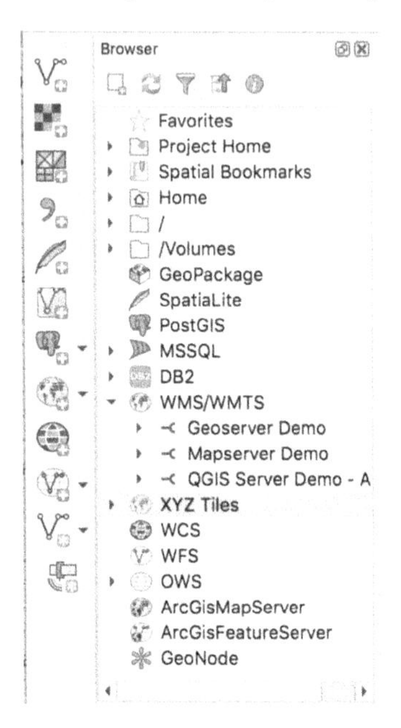

Along the top of your QGIS screen are the menus and toolbars. The toolbars contain icons arranged by function (for example, help, project, digitizing). The icons are used to access QGIS functions. You will probably have less than mine, as I have added several Python plugins.

2.5 QGIS Icons

Hovering your cursor over any icon will bring up a description. Try it. You can choose which of these are viewed, and can move groups of them by clicking on the vertical gray areas at the left of each group of icons . The double » icon in the toolbar means that there are additional icons in that group that are not visible. Click the » to view them.

You can move a toolbar to different areas (but not individual icons) by grabbing the vertical gray bars and dragging it. They can be moved along the right, left, and bottom sides of the window. Some icons may be grayed out if not active. Also, all the functions in the icons are duplicated in the main menu at top, so you can get rid of the any or all of the icons after you become more familiar with QGIS using the `View->Toolbars` menu.

Move over the icons and try moving a toolbar to the right or bottom of your screen.

2.6 Adding Data To QGIS

There are several ways to add GIS data to your new project. The easiest is to browse your data and simply drag any valid vector, raster, or other dataset from the *Browser Panel* into the empty map space at the right. It will automatically load it. You can do the same using the file manager on your computer. Browse a directory with GIS data files, then drag and drop one or more files onto the map area. You can also use the `Layer->Add Layer` menu and you will see all of the options to add different data types:

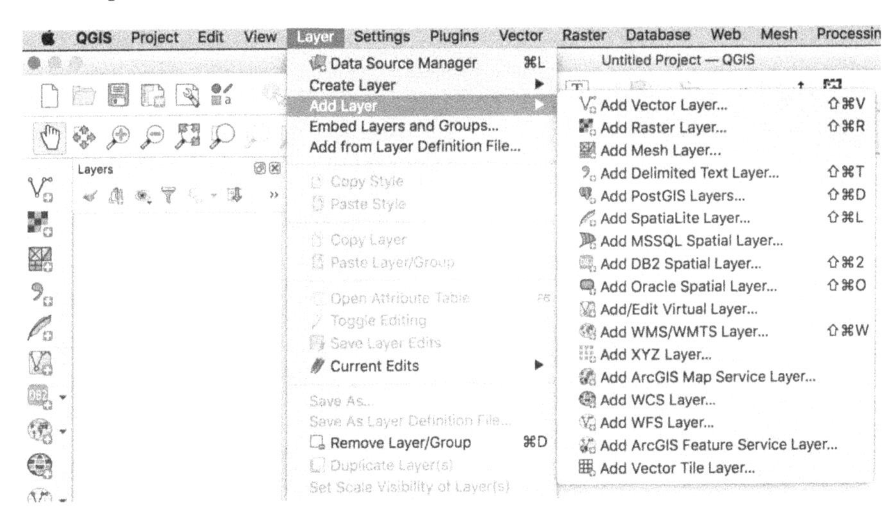

Another way to add data is using the series of icons on the left side of the QGIS screen. If they aren't visible, enable the *Manage Layers* toolbar using the `View->Toolbars` menu.

If you hover over these you will see that the top one is to add vector data, then raster data, etc. Simply click on the type of data you want to add, and it will take you to the appropriate dialog box. Or you can click on the *Open Data Source Manager* icon to open a new window:

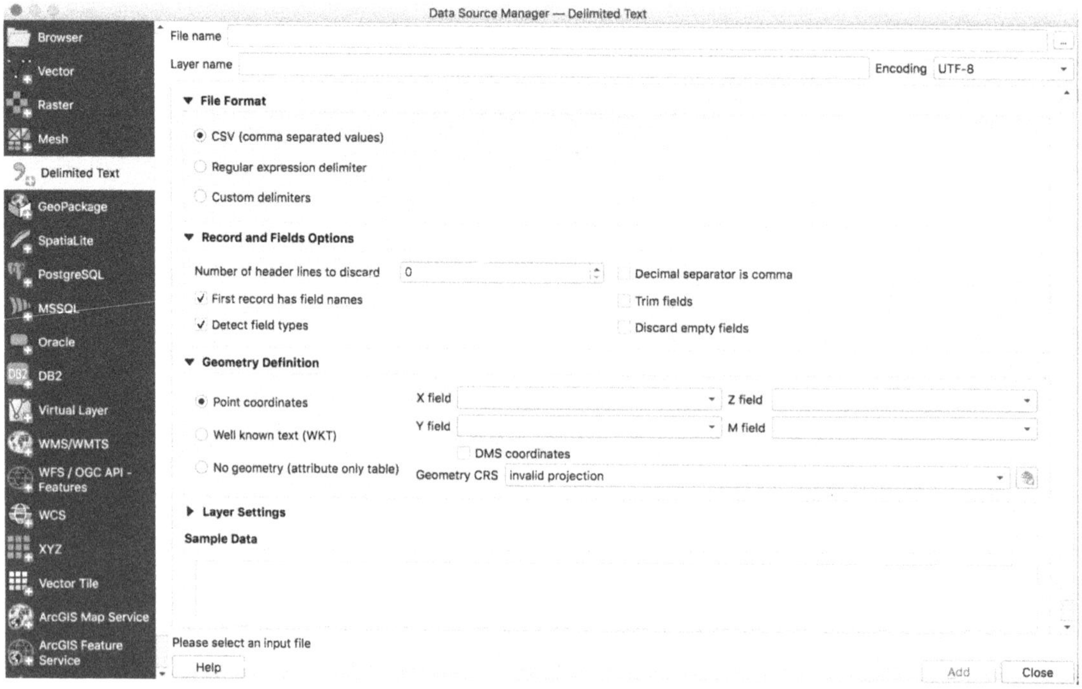

This has a list of options down the left side that lets you add vector, raster, database, Oracle, WMS, etc. files. You can also simply click on any valid data file directly from your files and drag it into the QGIS display window. This is the simplest way to add vector, raster, etc. data to QGIS. Give it a try. For shapefiles or GeoPackages, you only need to drag the .shp file or GeoPackage file into the map space at right.

Go to the directory your tutorial GIS data is stored, and in the Cape GeoPackage folder, drag `scott.gpkg` into your map space, the large, open white area on the right of the QGIS window:

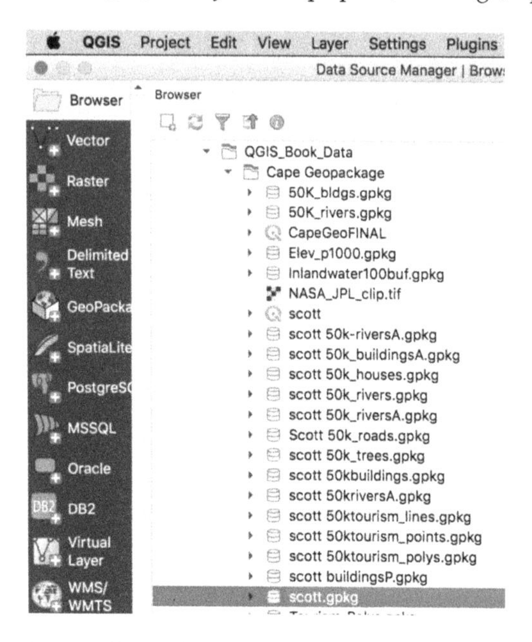

You will see this dialog box. It shows you the individual files in that GeoPackage.

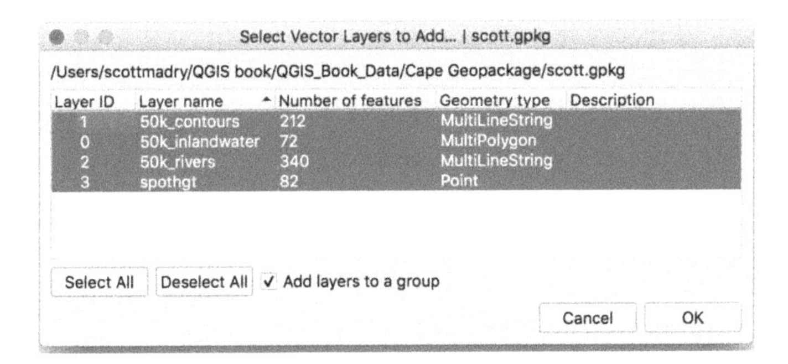

Click on *Select All*, then *OK* to load the four layers:

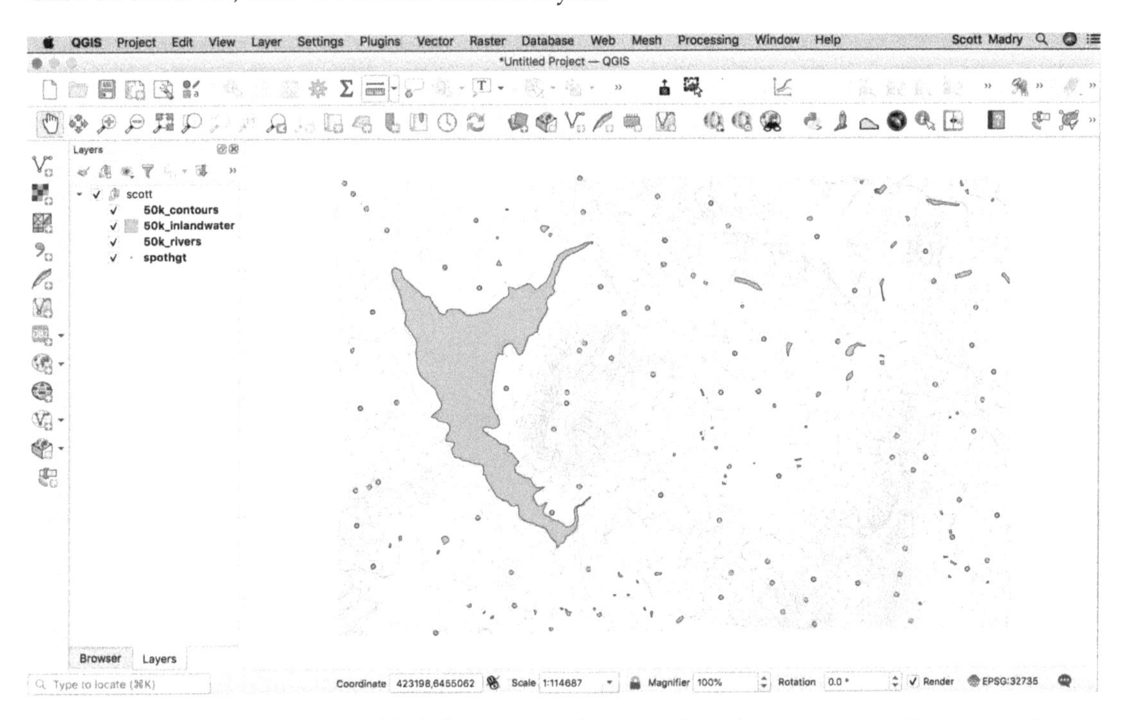

You will see that you have added four vector layers: elevation contours, rivers, a polygon inland waters layer (lakes), and a point elevation layer. These are listed in your *Layers* panel at left. You can click them off and on by checking or unchecking the name in the *Layers* panel. You can resize the panel by grabbing the gray bar at the right.

Now we will add a 1:50,000 topographic map that has been scanned and georeferenced (Kommandodriftdam # 3226AA, dated 1996, produced by the Republic of South Africa, Chief Directorate: Surveys and Mapping `http://www.ngi.gov.za`). Click on the *Open Raster* icon at the top left, (or use the `Layer->Add Raster Layer` menu), and click on the icon at right to navigate to the `QGIS_data_files` `Cape_dataUTM` folder.

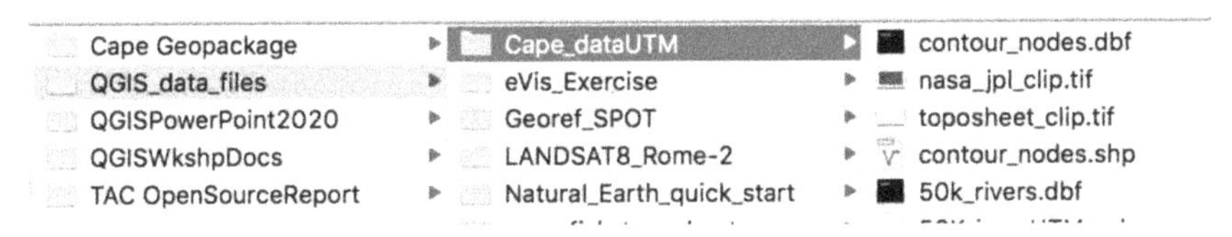

Now click on the file named `toposheet_clip.tif`, and it should look something like this:

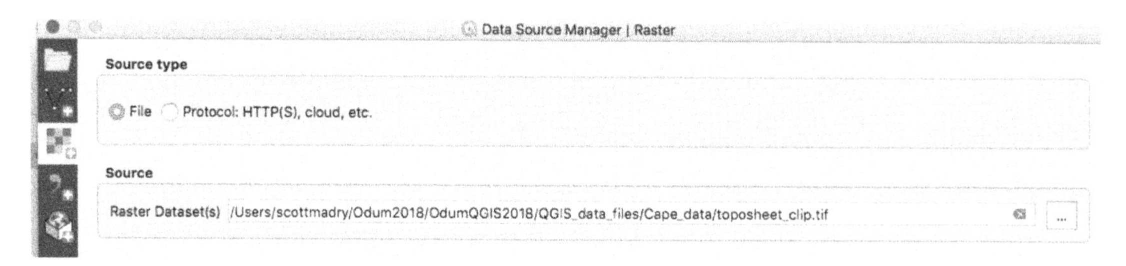

Click *Add* at the bottom right, and your raster map will be added as shown below:

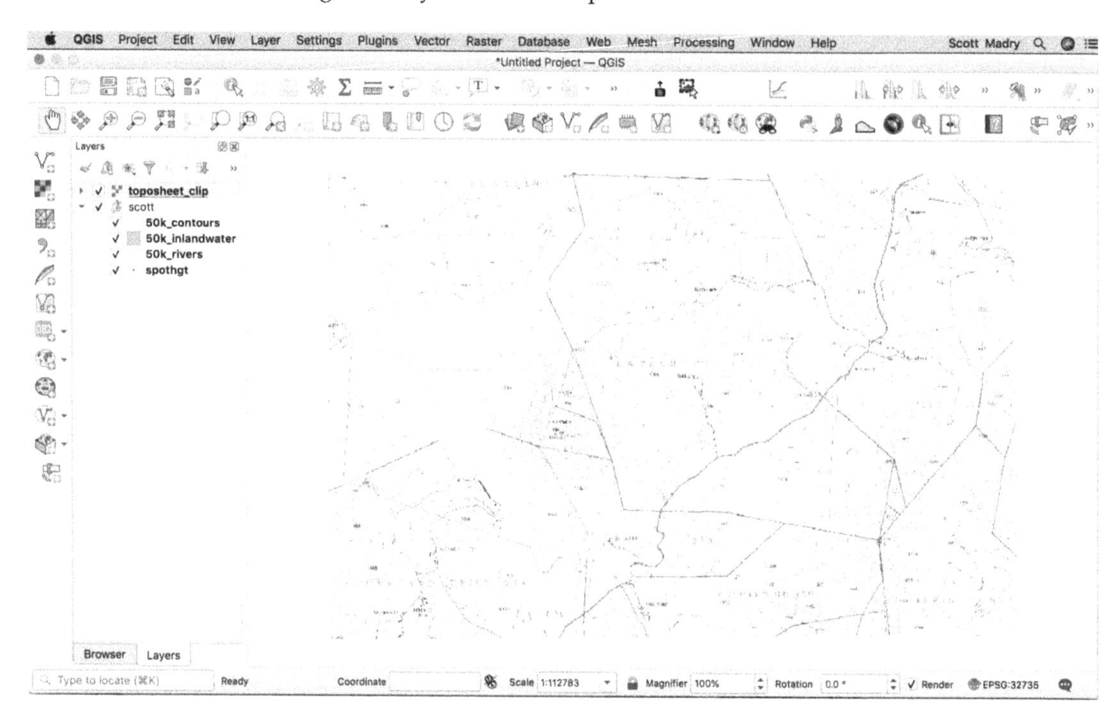

This is a 1;50,000 South African topographic map, and it was from this map that the other vector GIS data were digitized in QGIS. The raster map, if loaded at the top of the *Layers* panel, is covering up the vectors, so click on the `toposheet_clip` name in the *Layers* panel at left (it will turn blue) and drag it to the bottom of the list—the other layers will now be visible.

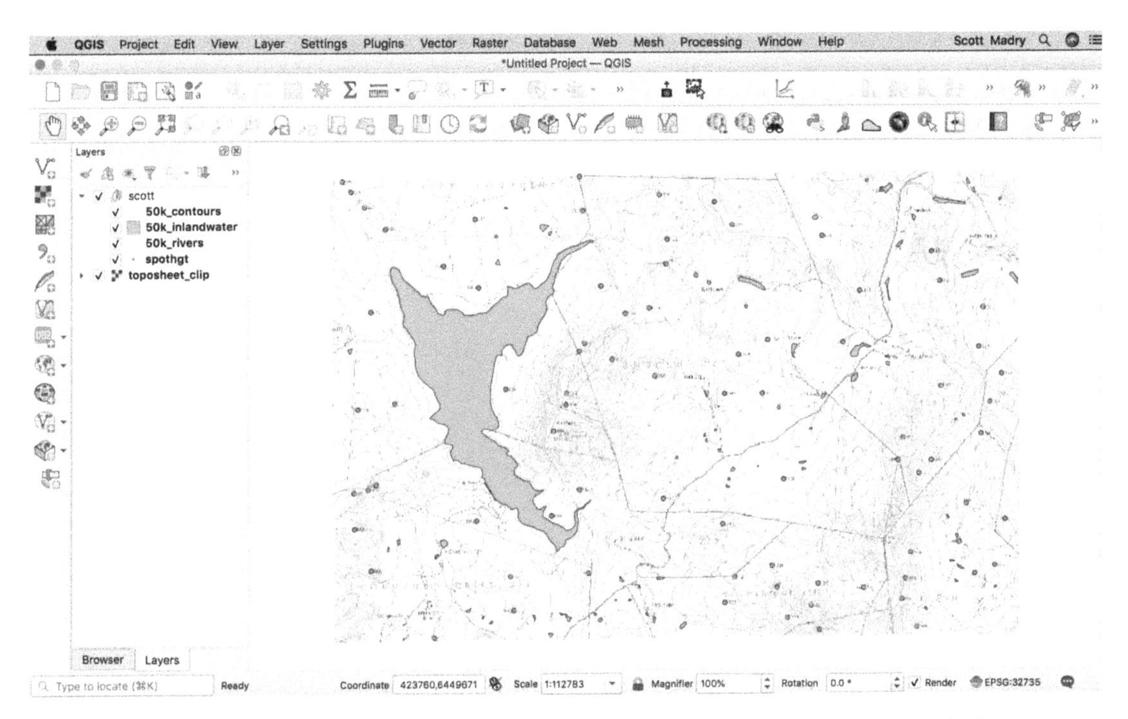

In general, we usually want to display point data on the top of our *Layers* panel, then vector line data, then polygon data, and put our raster data on the bottom. You can reorder these by just clicking on one and moving it up or down. Go ahead and do this. Note that you can also turn an individual layer on and off by clicking on the check in the *Layers* panel.

Try it. Also note that when you move your mouse over the map, it will show the coordinates, and also the scale, rotation, and other info.

You can have more than one data window open. Use the `View->New Map View` to open a new map window.

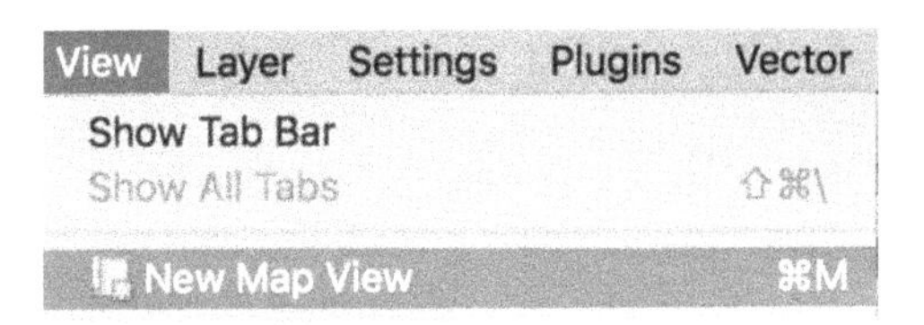

and you will see a second map window appear.

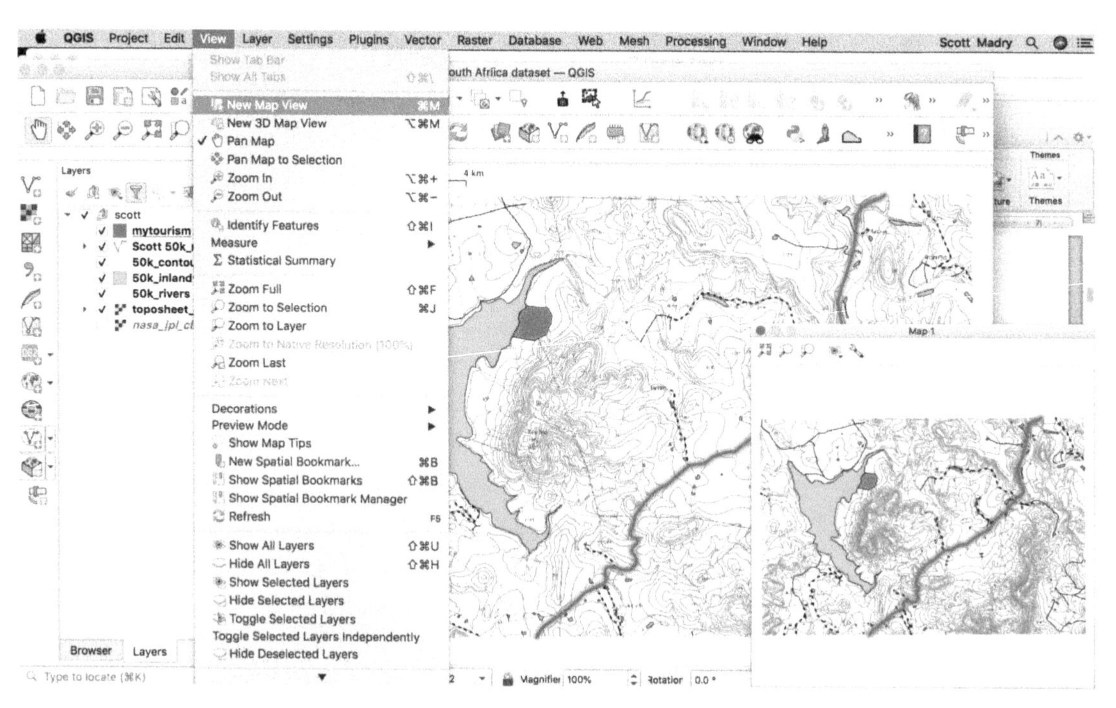

Note that this new window has additional icons and pull down items. The right icon opens a new dialog box to control the scale, synchronization, etc.

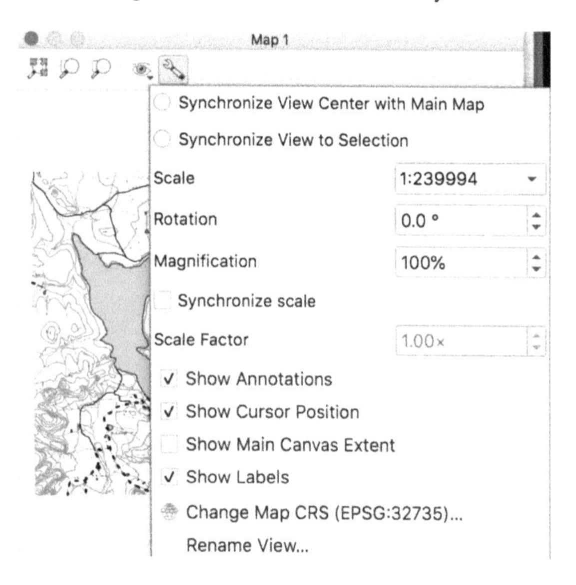

2.7 The Layers Panel

Note that at the top of the *Layers* panel there is a set of icons:

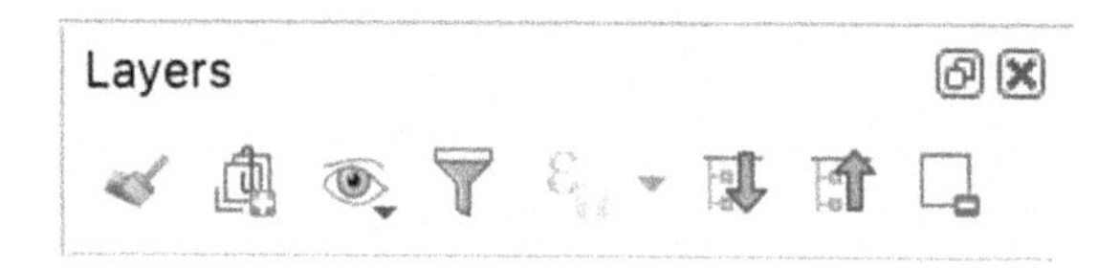

Hover your mouse over these to see what they do. They allow you to select and filer layers to be displayed. The left icon opens the *Layers Styling* panel, which allows you to choose different colors, line thicknesses, etc. for your vector data. The third icon lets you manage your map themes, as shown below:

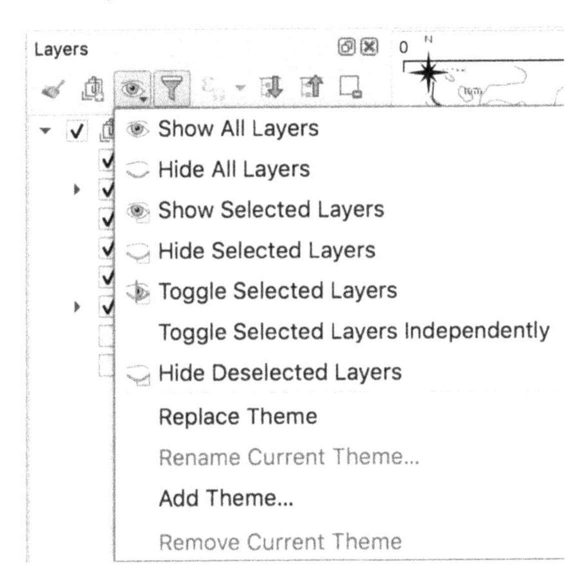

Click on the left icon for layer styling:

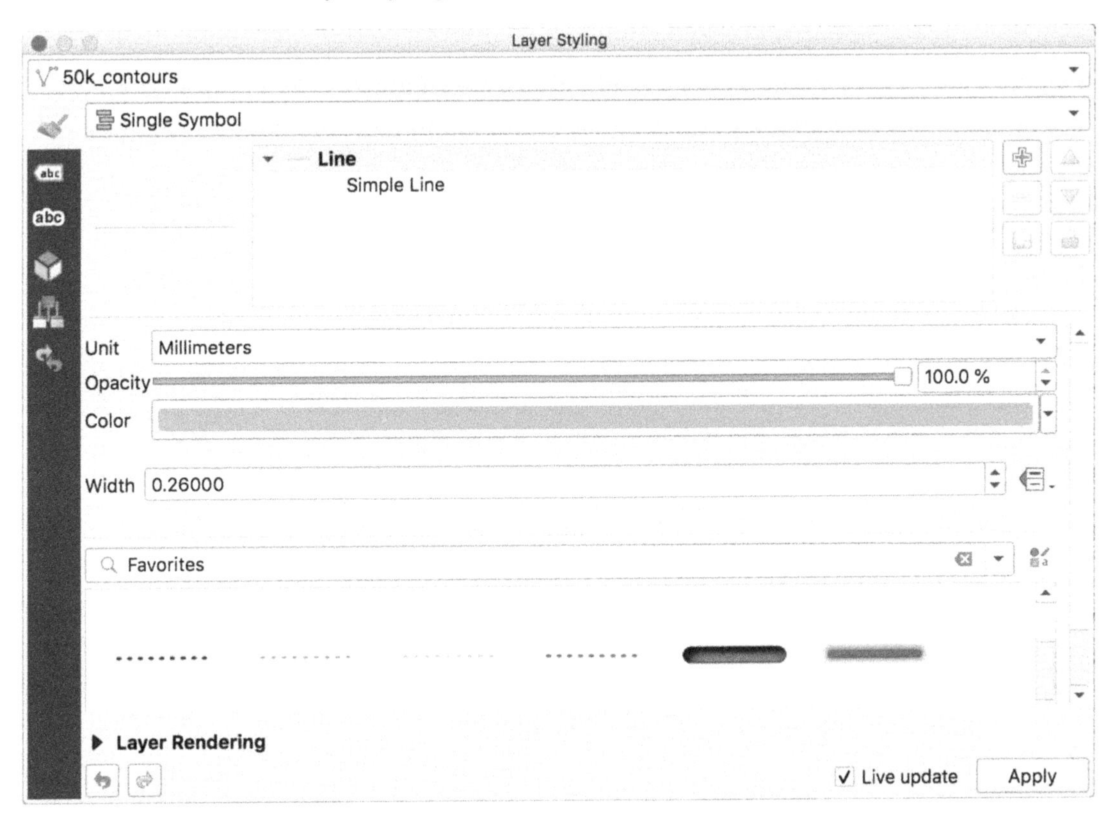

Now right-click on the name `toposheet_clip` in the *Layers* panel at left.

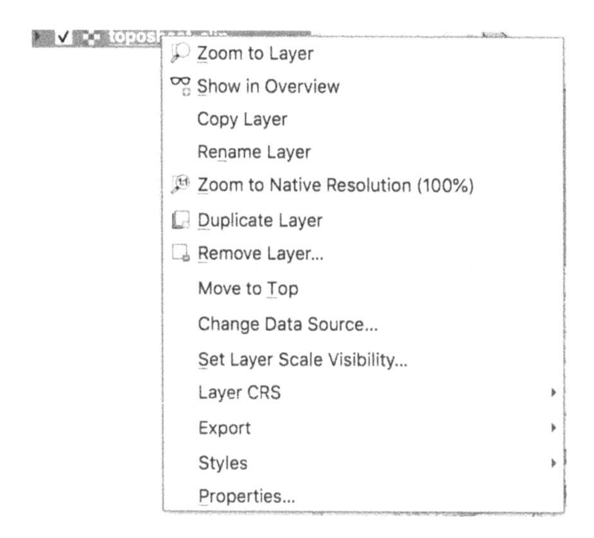

This is one of the most common groups of QGIS functions. Here you can zoom to the layer extent, zoom to the native raster resolution, and set the Coordinate Reference System (CRS), rename layers, export features, and other common functions. Right-click on one of the vector files and again, you will see a slightly different group of options:

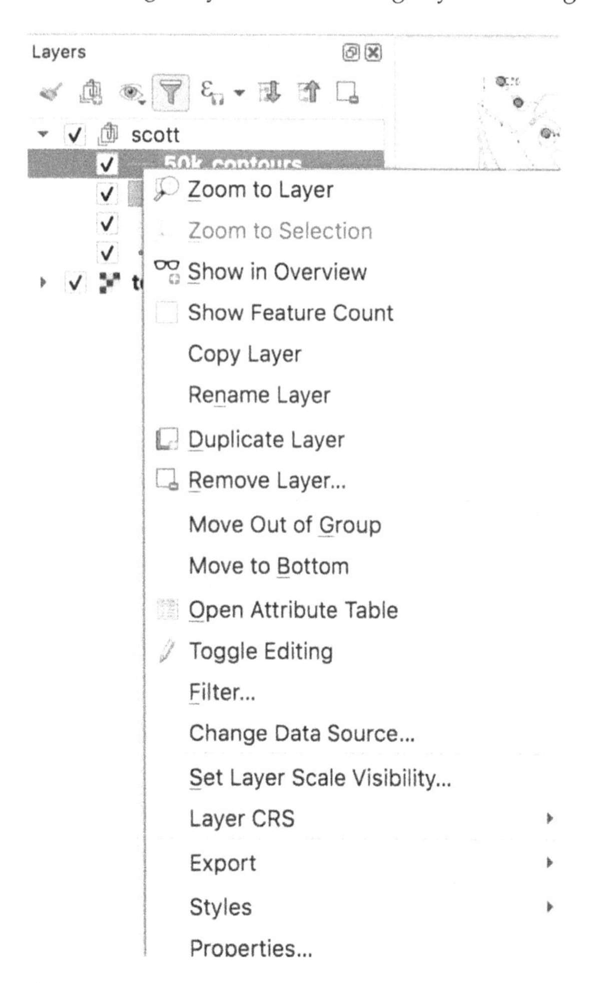

With vector data, we can do many of the most common functions, which we will review shortly. Now that we have the beginnings of a working QGIS database loaded, you will see at the bottom that as you move your mouse in the data viewer the coordinates are shown in a box below, and that you can choose what scale to view your data or type in the scale you want. It also contains, at right, the current Coordinate Reference System (CRS) for the project, displaying the EPSG code (a common code of coordinate systems), and allows you to change this by clicking the right-most button. You can zoom by changing the Magnifier setting, or rotate off North as well. Try changing the scale to 1:50,000. You can zoom in and out using your mouse scroll wheel or trackpad.

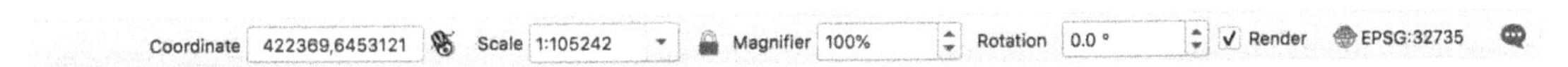

Click on the ESPG box at right and it will open the tab that shows the Coordinate Reference System for this project:

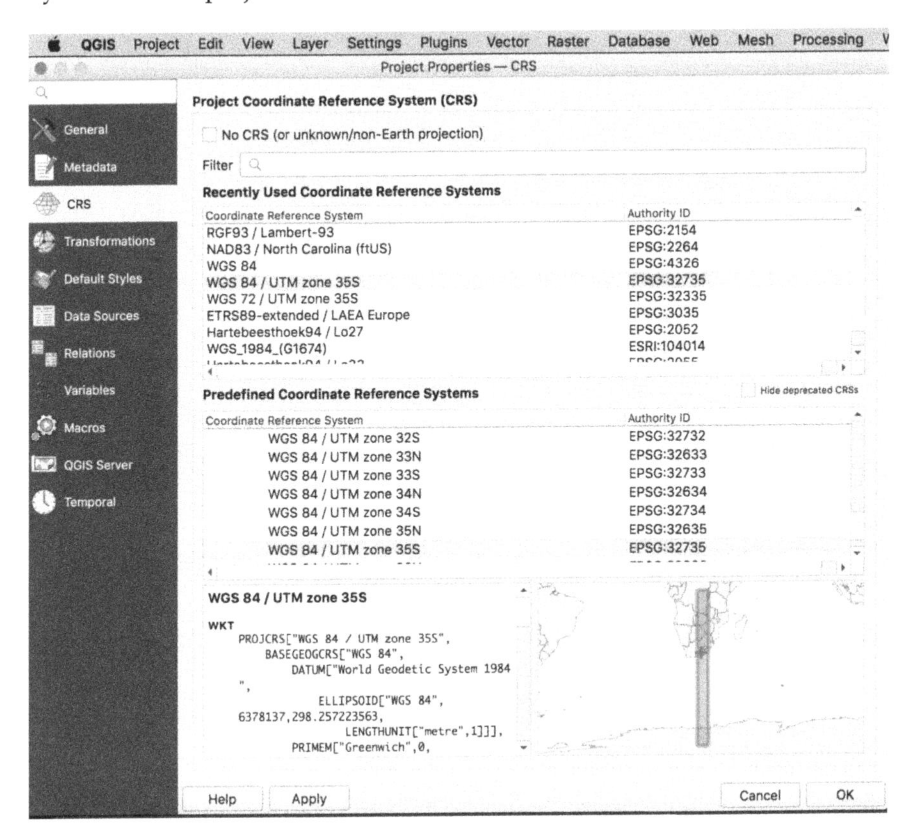

You can see that we are using the WGS 84/Universal Transverse Mercator zone 35 South coordinate system for this database.

Now look at the icons along the top.

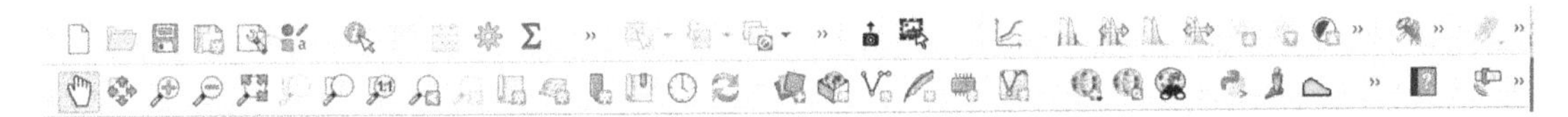

These include the tools for manipulating, adding, and displaying data. Hovering the mouse over each will bring up a description. We will review these directly.

Above the icons are the pull-down menus that provide access to all QGIS functions and capabilities for saving, displaying and analyzing data.

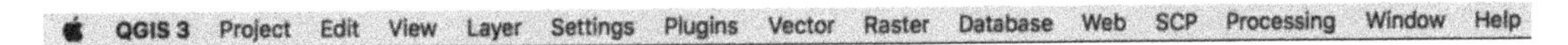

Look under the various options that are available by clicking on one of the menu options. We will cover all of these. All the functions in the icons are also duplicated in the menus as well. I probably have some you don't, as I have added several plugins. You can turn off groups of the icons by going to `View->Toolbars` and unchecking them:

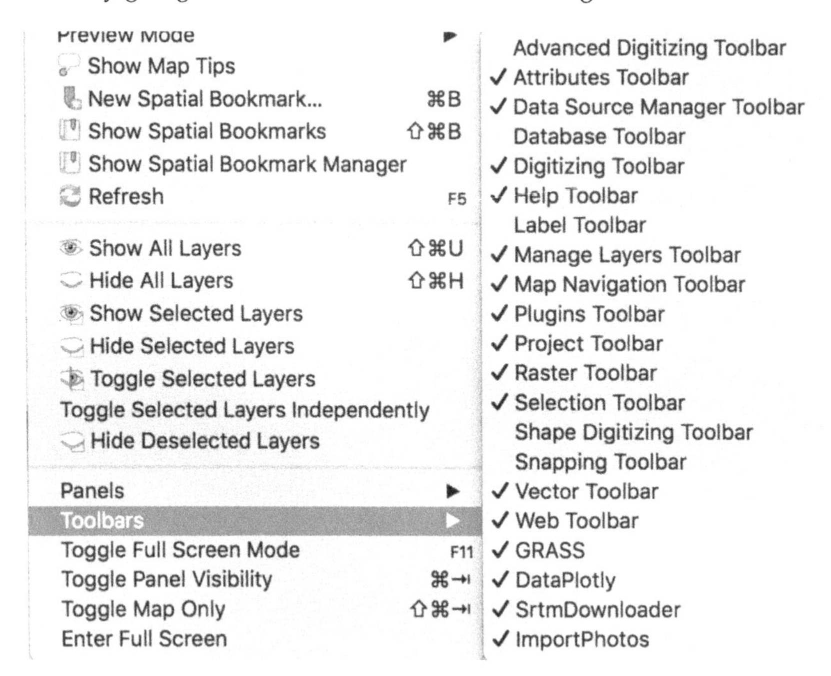

2.8 Adding Vector Data

Now add some more vector files to the map using the *Add Vector* icon at the left of the QGIS screen, or from the `Layer->Add Layers->Add Vector Layer` menu.

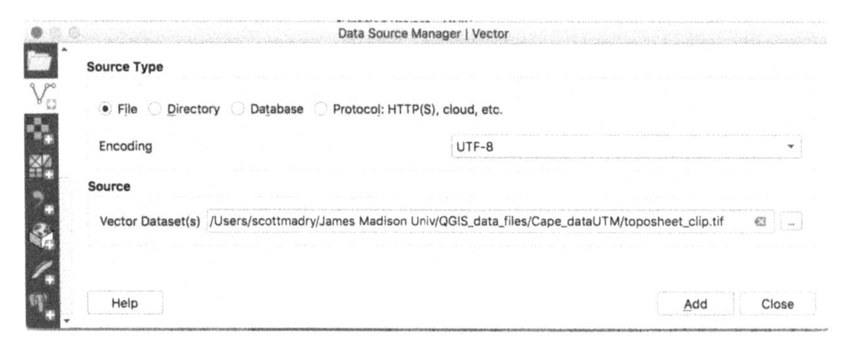

Click on the three dots to the right of the *Vector Dataset(s)* field and navigate to the same Cape GeoPackage (it should take you right there since we just used it) to add a vector shapefile. Note that you can change which file types will appear by clicking the white box at the center bottom. It will be set on *All files*, but change this to *ESRI Shapefiles*, and then you will only see the .shp files highlighted. You can see we can import many different types of vector files into QGIS.

Double-click on `Scott 50k_roads.gpkg` to choose it and return to the Data Source Manager dialog. Note that the source path is now filled out. Click *Add* at the bottom and close the window:

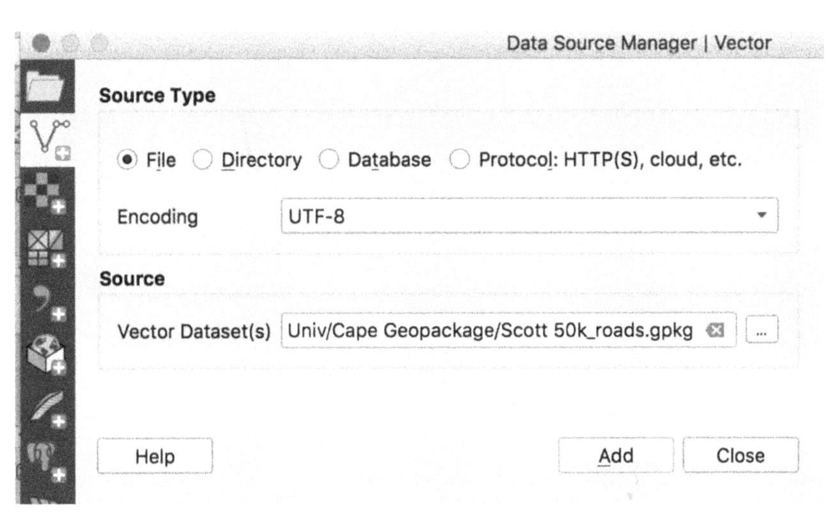

Your new vector roads layer will now appear in your *Layers* panel. This layer uses the WGS84 CRS—we'll deal with that in a minute. Remember that you can also simply grab and drag any recognized file over to the display and it will load automatically. Shapefiles actually consist of several files, so always drag only the .shp file.

2.9 The Layers Panel

Your *Layers* panel will now look something like this:

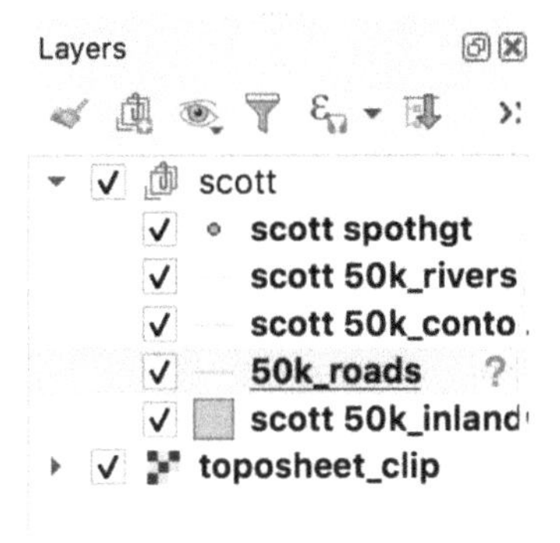

Note the *?* by the roads file name. Click on it and you will see that its CRS is not defined. Since we previously stated it was WGS84, click on *WGS84* and click *OK*.

QGIS assigns a random color to new vector files, and you may not be able to clearly see the roads layer on your screen, so double-click on the small *line* between the check and the name of the file: ✓ — **Scott 50k_roads**

This will take you to the *Symbology* panel of the *Layer Properties* for this layer:

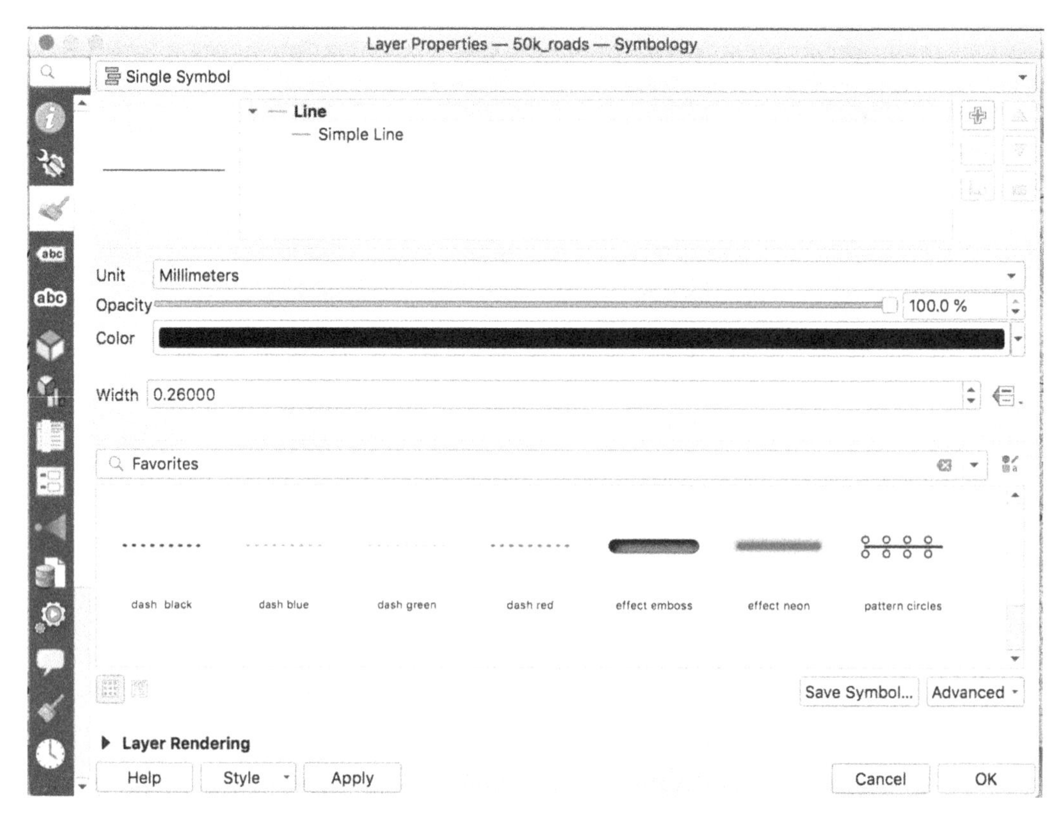

You can also get there by right-clicking on the layer name in the *Layers* panel, and choosing `Properties` from the menu. Click on the `Symbology` tab to change the color, line width, etc. of vector point, line, and polygon data. Note that you have several "canned" options at the bottom to choose from. If you click on the small down arrow to the right of the *Color* bar, you can change the color of the roads vectors.

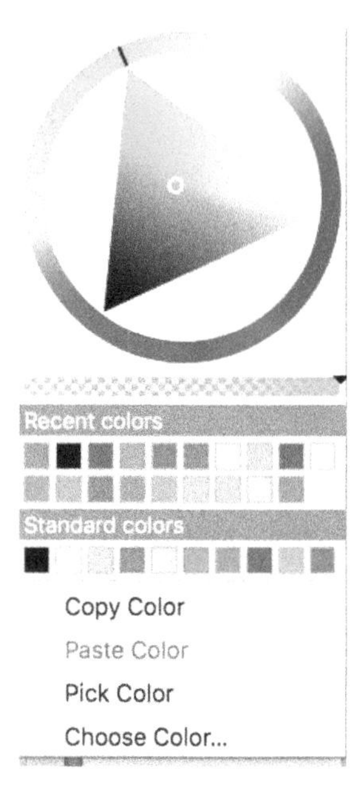

You can also change the line width, opacity, and other aspects, click *OK* on the *Properties* panel. Note that there are several "canned" vector line options displayed at the bottom that you can choose from.

Your roads layer will now be more visible. Using the symbology options, you have complete control over how vector points, lines, and polygons are presented:

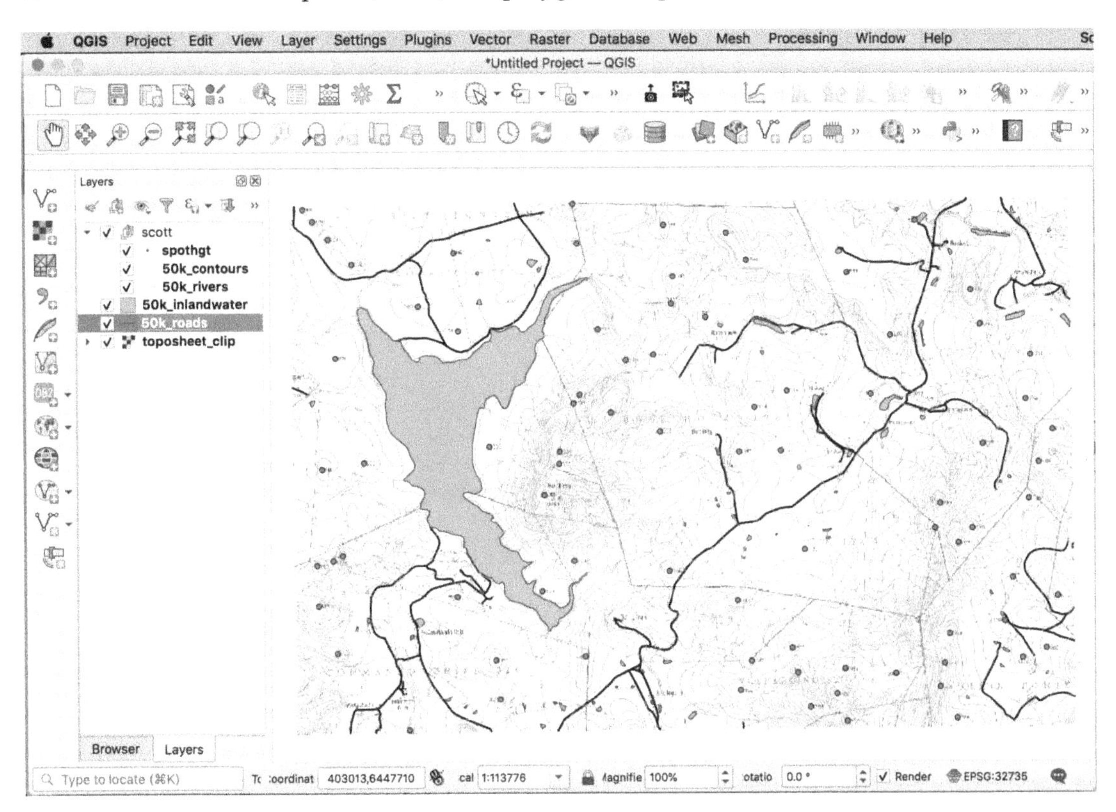

Note there may be a warning in yellow stating that the CRS was not defined in some layers. QGIS will automatically present them in the CRS of the current project.

Continuing to work with the roads layer, right-click on it, then click on `Open Attribute Table`:

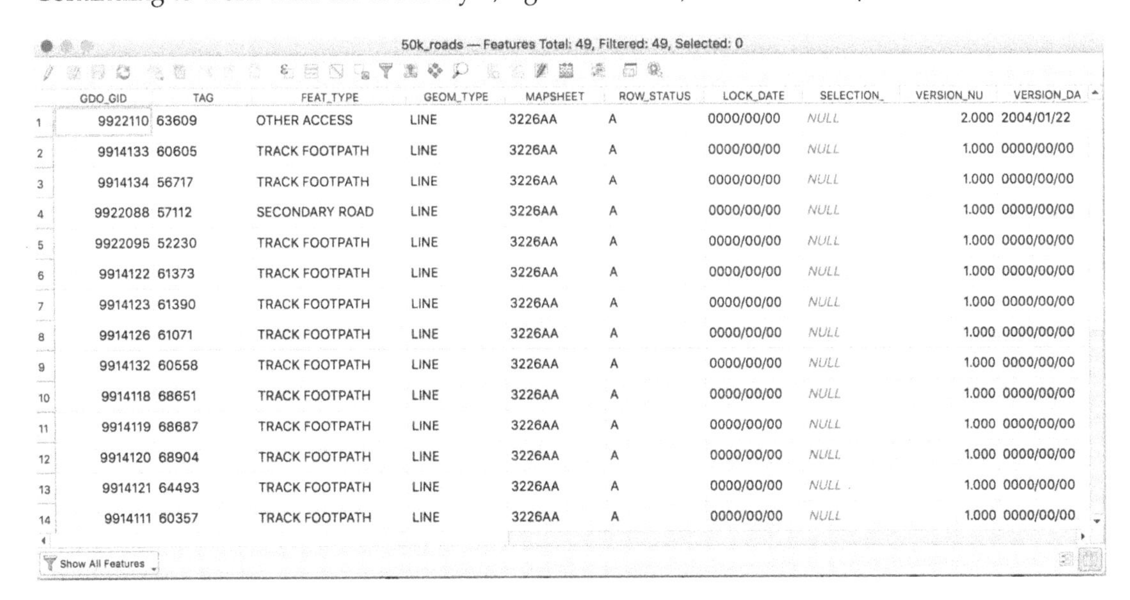

50k_roads — Features Total: 49, Filtered: 49, Selected: 0

	GDO_GID	TAG	FEAT_TYPE	GEOM_TYPE	MAPSHEET	ROW_STATUS	LOCK_DATE	SELECTION_	VERSION_NU	VERSION_DA
1	9922110	63609	OTHER ACCESS	LINE	3226AA	A	0000/00/00	NULL	2.000	2004/01/22
2	9914133	60605	TRACK FOOTPATH	LINE	3226AA	A	0000/00/00	NULL	1.000	0000/00/00
3	9914134	56717	TRACK FOOTPATH	LINE	3226AA	A	0000/00/00	NULL	1.000	0000/00/00
4	9922088	57112	SECONDARY ROAD	LINE	3226AA	A	0000/00/00	NULL	1.000	0000/00/00
5	9922095	52230	TRACK FOOTPATH	LINE	3226AA	A	0000/00/00	NULL	1.000	0000/00/00
6	9914122	61373	TRACK FOOTPATH	LINE	3226AA	A	0000/00/00	NULL	1.000	0000/00/00
7	9914123	61390	TRACK FOOTPATH	LINE	3226AA	A	0000/00/00	NULL	1.000	0000/00/00
8	9914126	61071	TRACK FOOTPATH	LINE	3226AA	A	0000/00/00	NULL	1.000	0000/00/00
9	9914132	60558	TRACK FOOTPATH	LINE	3226AA	A	0000/00/00	NULL	1.000	0000/00/00
10	9914118	68651	TRACK FOOTPATH	LINE	3226AA	A	0000/00/00	NULL	1.000	0000/00/00
11	9914119	68687	TRACK FOOTPATH	LINE	3226AA	A	0000/00/00	NULL	1.000	0000/00/00
12	9914120	68904	TRACK FOOTPATH	LINE	3226AA	A	0000/00/00	NULL	1.000	0000/00/00
13	9914121	64493	TRACK FOOTPATH	LINE	3226AA	A	0000/00/00	NULL	1.000	0000/00/00
14	9914111	60357	TRACK FOOTPATH	LINE	3226AA	A	0000/00/00	NULL	1.000	0000/00/00

Show All Features

More on the attribute tables and databases a bit later, but you can see that there is a category called `feat_type` that includes the types of roads. Close the attribute table, and again right-click on your roads layer and choose `Properties->Symbology`. At the very top, change single symbol to *Categorized*, and for *Value*, use `FEAT_TYPE`. Then click on *Classify*, and you will see the types of roads. Click *OK*, and now look in the *Layers* panel and you will see that you have categorized your roads based on type.

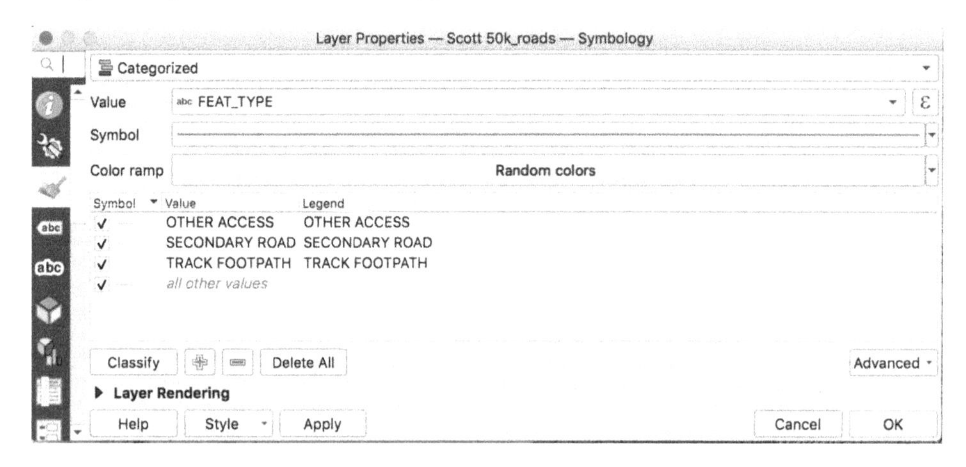

These can now be assigned different colors, and can be turned on and off individually, and be symbolized in different ways.

Lets explore the *Layers* panel at left, where you have all the loaded layers listed. You can toggle the display of each layer by clicking on the check marks in the *Layers* panel.

scott spothgt

You will see this, where the raster map is shown with the various vector files displayed as well:

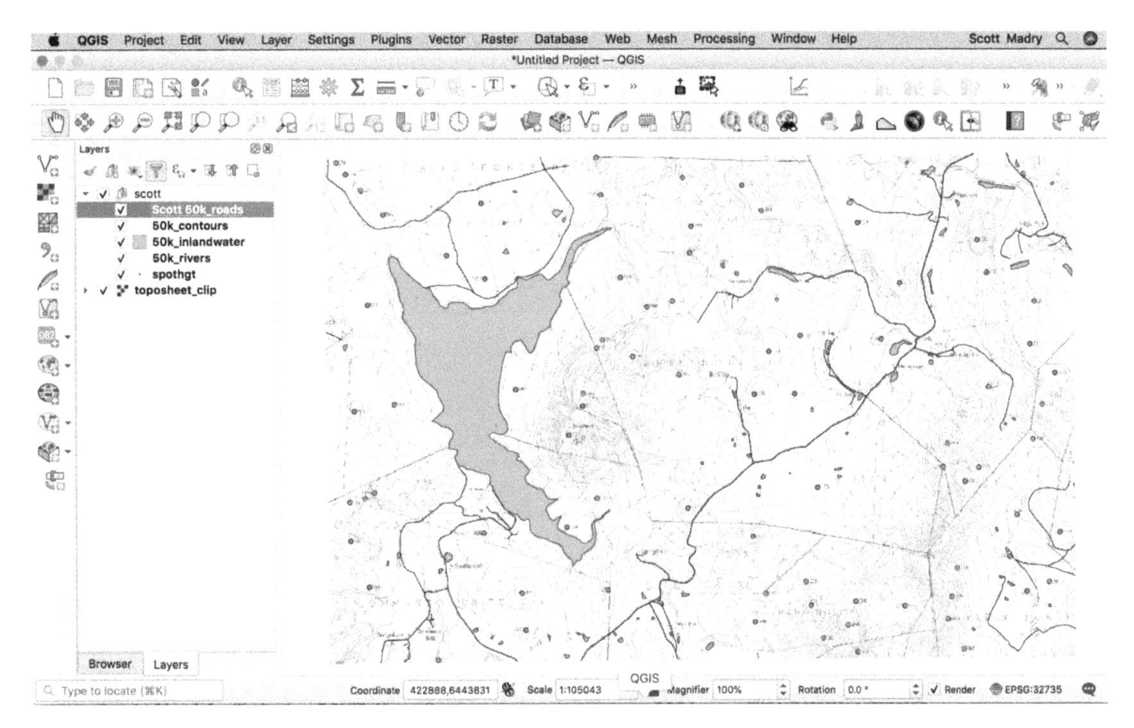

The order of files (top to bottom) in the *Layers* panel at left is important, as raster layers will mask others if put on top. Leave your raster layer `toposheet_clip` at the bottom. Move layers up and down by left-clicking on the name and dragging it up or down the list. It is good practice to put point layers on top, followed by line layers, polygon layers, and raster files at the bottom. You can turn layers on and off by clicking the check at left. Try it.

Remember that you can also add files to your project by simply navigating in your *Browser* panel to a Geopackage, shapefile, or other file, and dragging it over to the map viewing area.

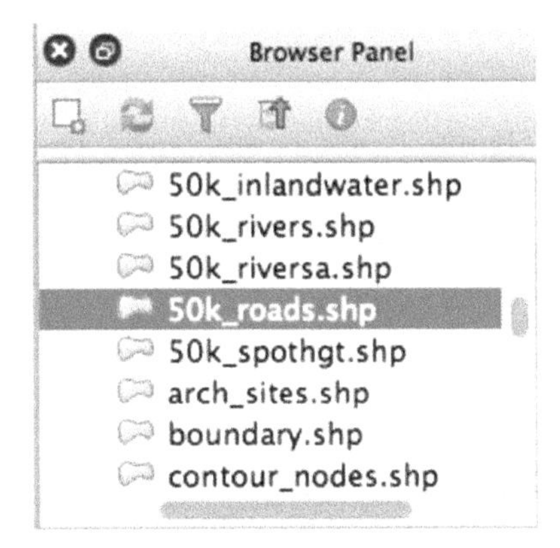

2.10 Layers Properties

Right-click on any of the file names in the *Browser* panel, and you will see a menu. Try the file named `50k_inlandwater`:

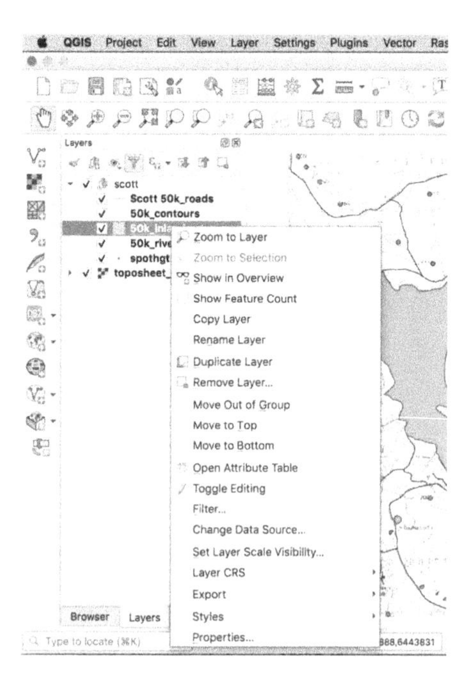

This menu lets you do many of the most commonly used QGIS functions, such as zoom to the extent of that specific layer (your GIS layers may not have the same extent), set the Coordinate Reference System (CRS), remove the layer from your project (this does not actually delete the file from your computer, it only removes it from this project), view the attribute table, edit the layer, or export it. You can duplicate a layer, meaning have two versions open at the same time to, for example, use different colors. It is not creating a new file, only displaying it twice in the project you have open. These are some of the most common sets of actions we do in QGIS. You open a layer and then right-click on it to use these options. Click on `Properties` at the very bottom, and you will open this dialog:

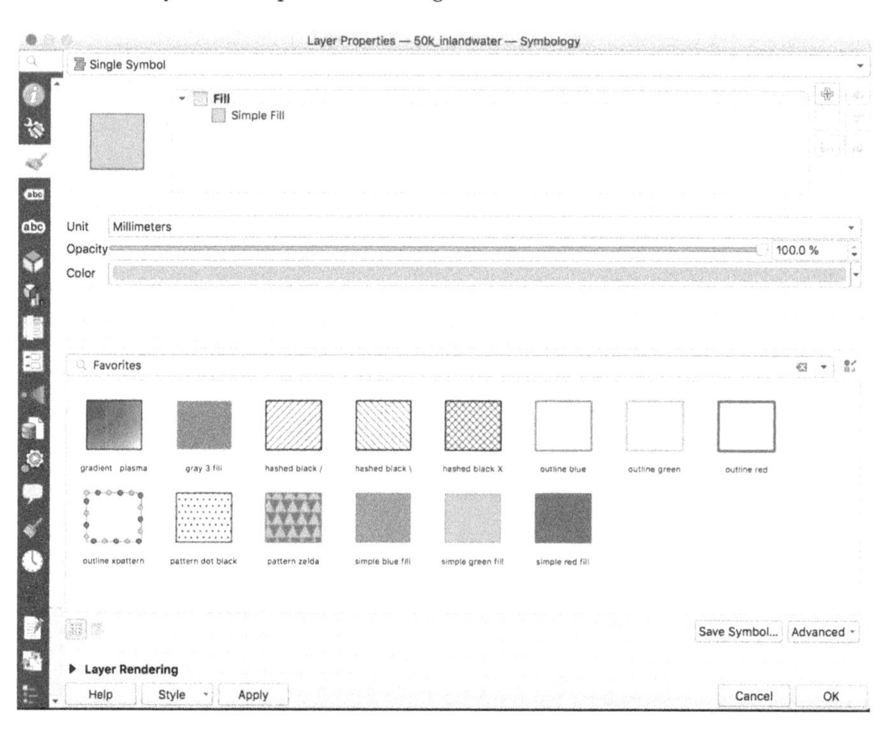

Note we are on the *Symbology* tab, as that was the last place we visited. This is where you change the symbology (color and width of vector layer, or change the symbols for point layer), add labels, record metadata, make files partially transparent, etc. We use this all the time. It will be slightly different for vector polygon, point, or raster data. Again, you have several "canned" polygon display and fill options to choose from at the bottom, and you can create your own as well. Hover your mouse over the tabs at left to see a description of each.

Click on the first tab (*General Information*) at the top left, to see the basic information about the layer:

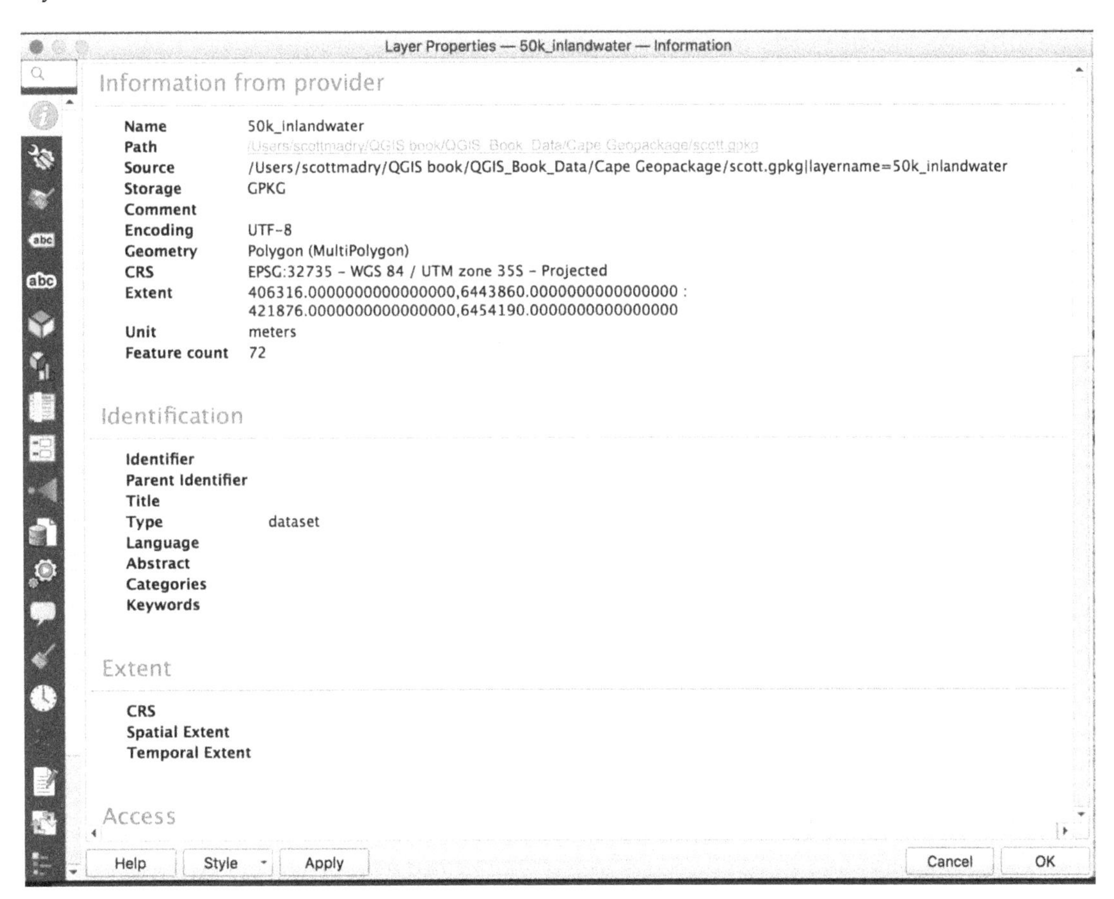

Scroll down to see all the information. Under the *Source* tab, you will see the geometry and CRS data:

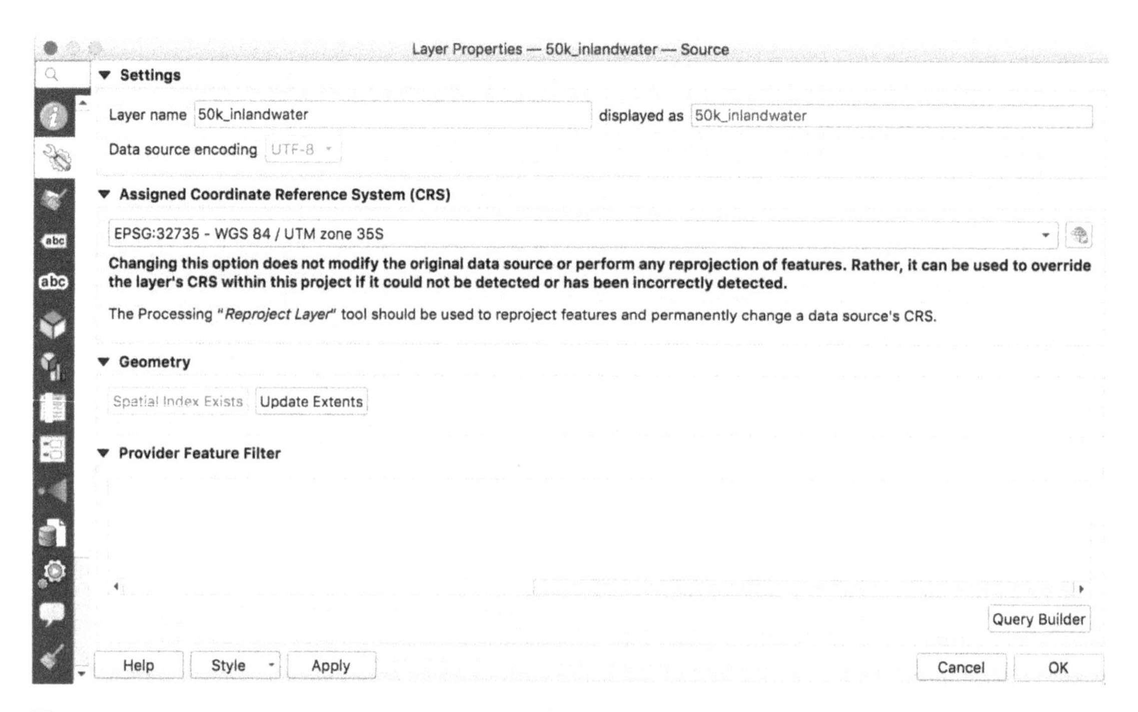

There are many more properties along the left-hand side that are important to learn. We will cover these later, but feel free to explore on your own. Now add another raster file, click on the *Open Raster* icon (or go to the `Layer` menu and click on `Add Raster Layer`) and navigate to the `Cape GeoPackage` folder and click on `NASA_JPL_clip.tiff` and *Open* down at the bottom right:

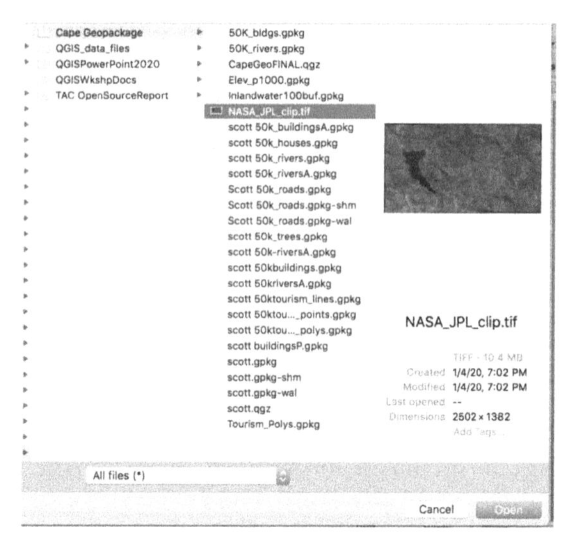

You may or may not see the satellite image. If you see a *?* by the name, click on it and you will see:

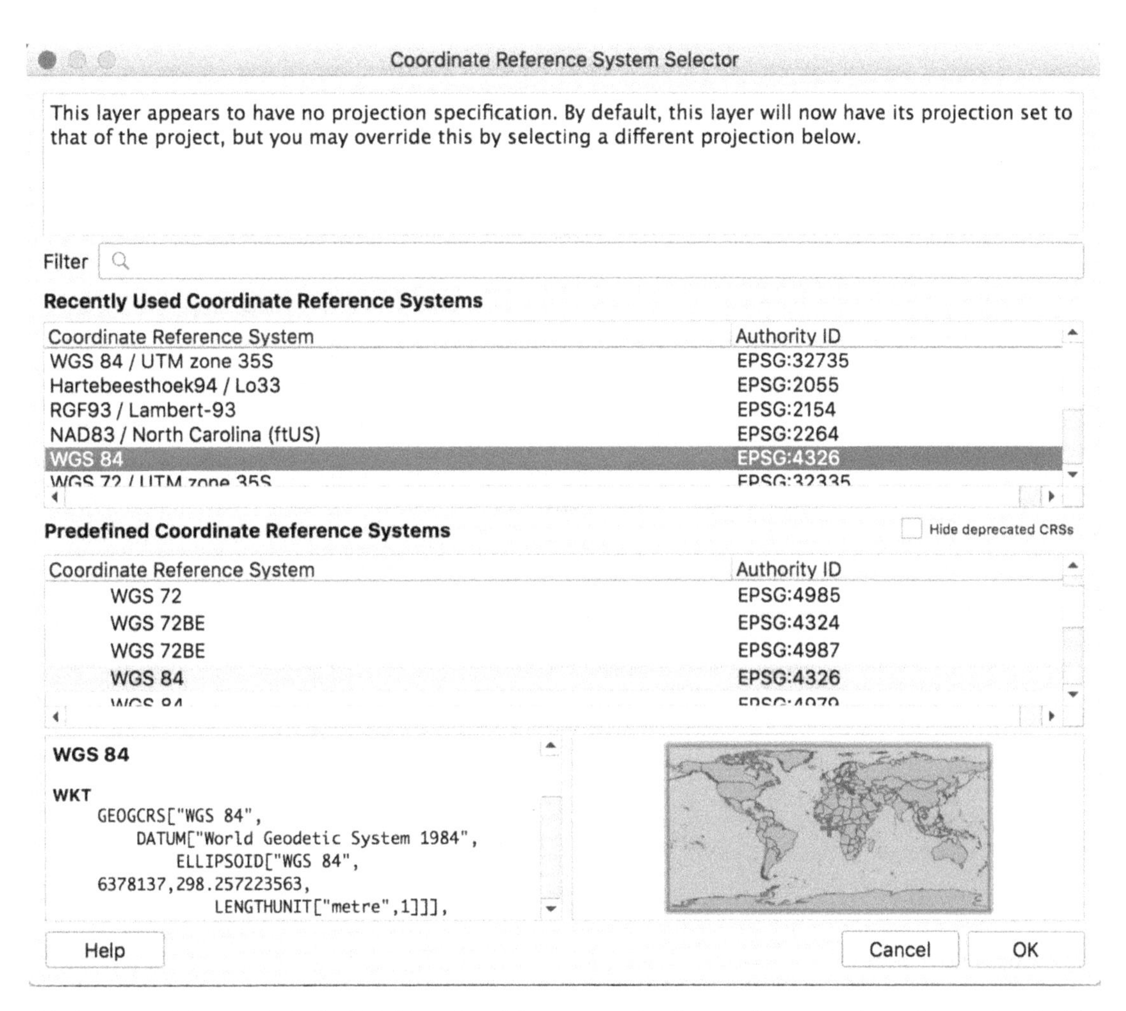

Click on WGS84 and click *OK*. You should now see it, depending on where it was loaded in the *Layers* panel. If you don't see it, drag it up to the top:

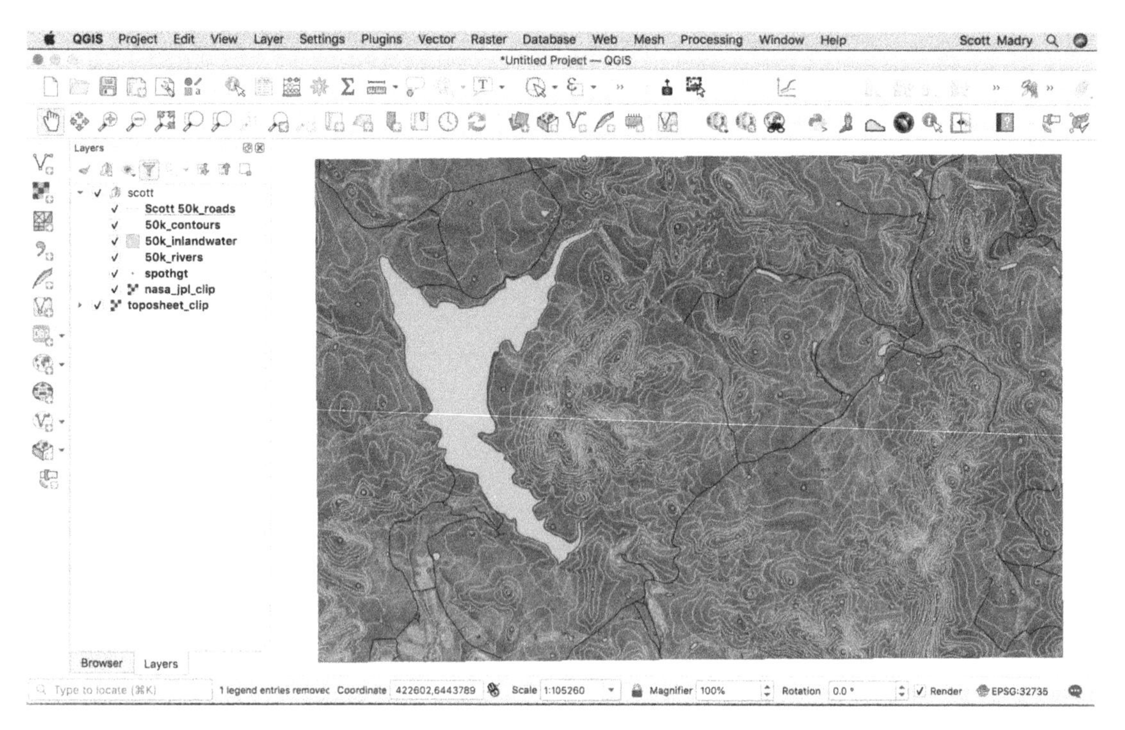

This is a NASA Satellite image of the area. Place it below the vector files in the *Layers* panel but above `toposheet_clip` so you can see the vectors, Right-click on the file name in the *Layers* panel at left click on `Stretch Using Current Extent`. This will do an automatic color stretch, which visually improves the image, as you see below.

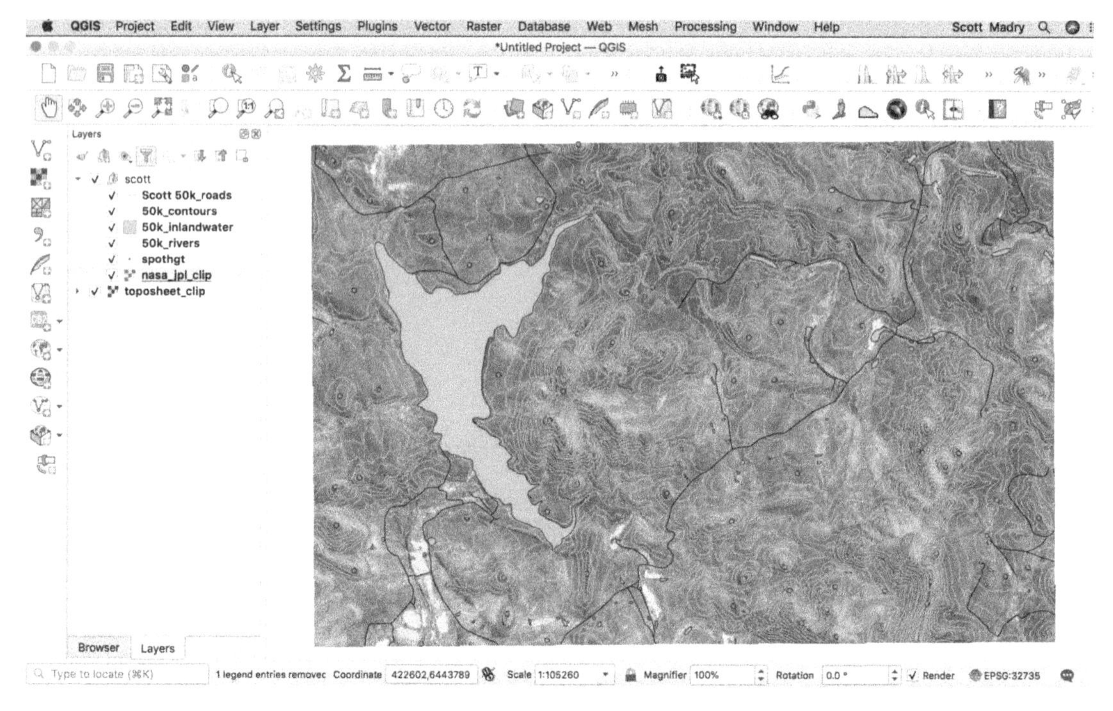

Raster files have additional options in the `Properties` menu, including `Zoom to Native Resolution 100%`, which shows you the actual spatial resolution of the raster data, as shown below:

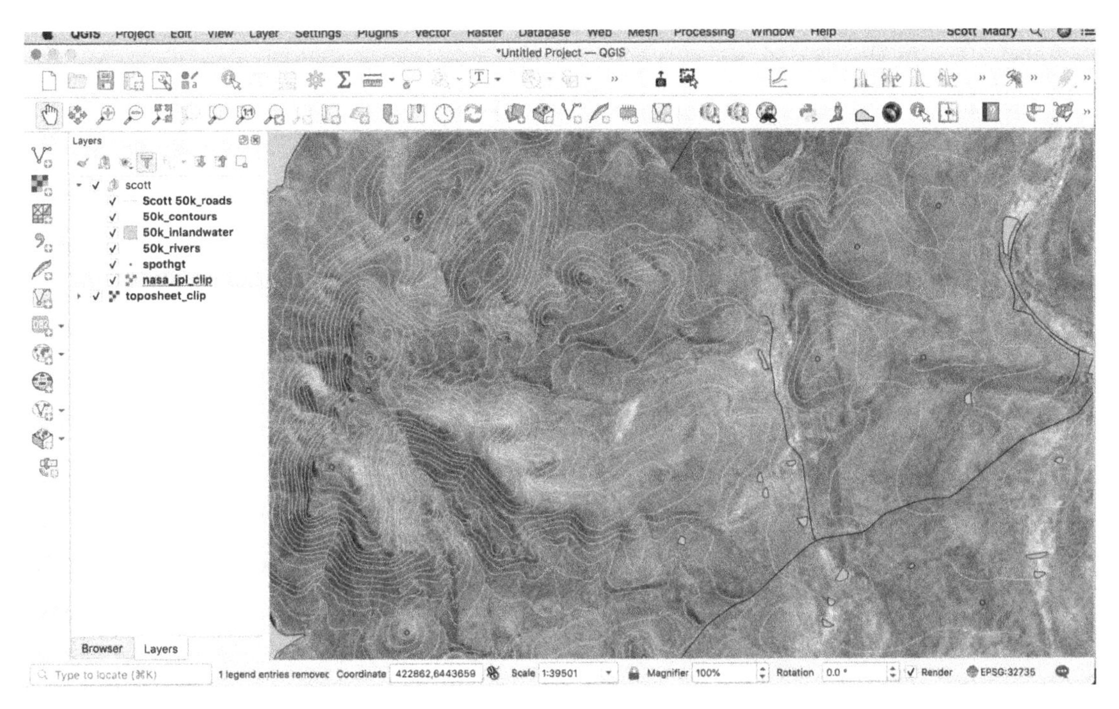

This is one of the main differences between vector and raster data. Vector data are just what it is, at any zoom or view. Raster data come with a specific spatial resolution, but this can be easily altered on-the-fly. Clicking `Zoom to Native Resolution 100%` resets the image to its true spatial resolution.

2.11 Map Interaction Using Icons

You can do this whenever you zoom into a small area. We clicked on `Zoom to Native Resolution 100%` to see the image zoomed in at its true spatial resolution. As we said, raster data, unlike vectors, have a native resolution, and this sets the display to it. Now right-click again and `Stretch Using Current Extent` and you will see it change the color table. Now `Zoom to Layer`, to see the entire image. Zoom in and wander around and look at the imagery, by clicking on the *Pan Map* icon (white hand) , and you can move the image. Zoom in and out using your mouse scroll wheel, or you can use the *Zoom In* and *Zoom Out* (plus and minus) icons in the toolbar.

Right-click on the name of the NASA image in the *Layers* panel, and right-click on `Properties`. You will see that raster files have some different properties from vector files.

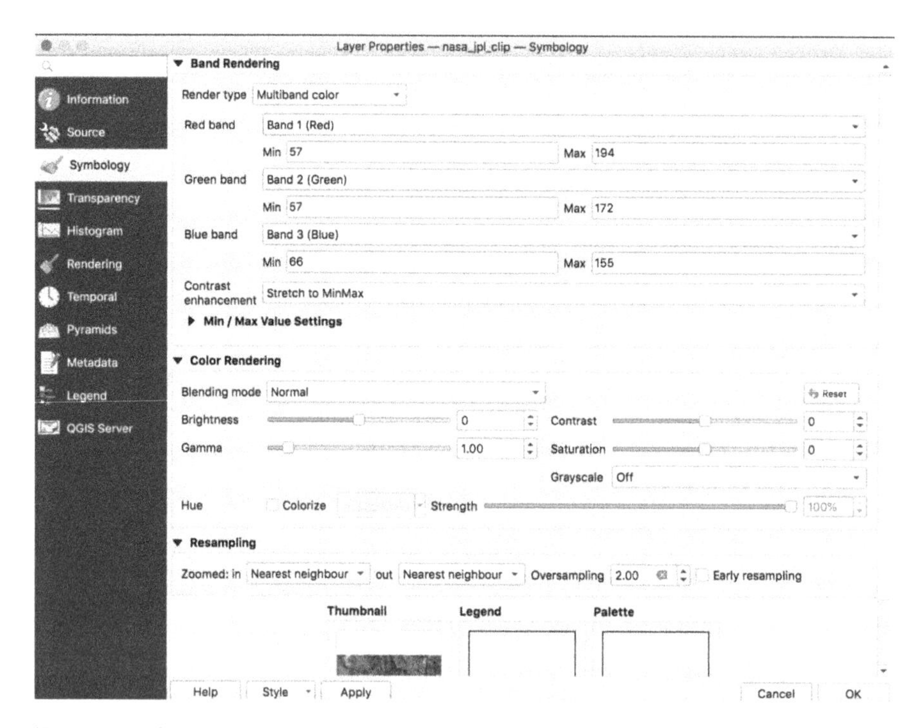

For example, you can see the individual band parameters in `Properties->Histogram`, and click on `Compute Histogram`. This shows you the RGB histogram for the satellite image.

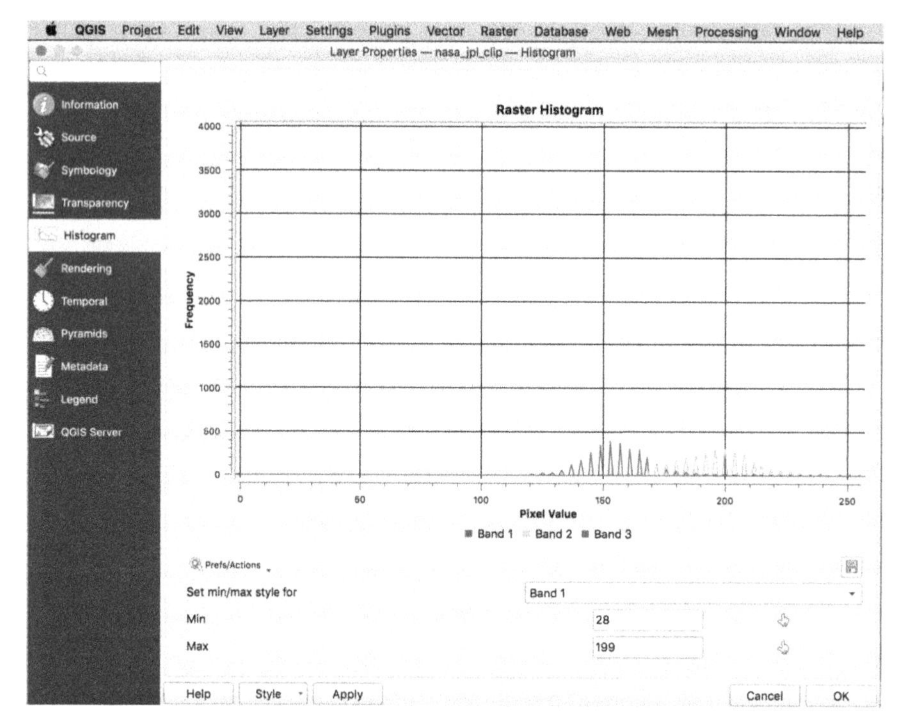

Zoom in and wander around and look at the imagery using the zoom and pan icons at the top left of your toolbar.

2.12 Setting File Transparency

Note that when you added the satellite image, it was automatically placed on the top of your data stack in the *Layers* panel, hiding all the vector data below it. Raster data "cover" all vector data below in the *Layers* panel, so grab the name of the image layer and drag it to the bottom of the stack. Now it is obscured by the topo map above it. You can click the `toposheet` layer off by unchecking the check mark. If you want to see both rasters at the same time, right-click on the `toposheet_clip` layer, go to `Properties->Transparency`, set the transparency to about 40%, and click *OK*. You will see this dialog box:

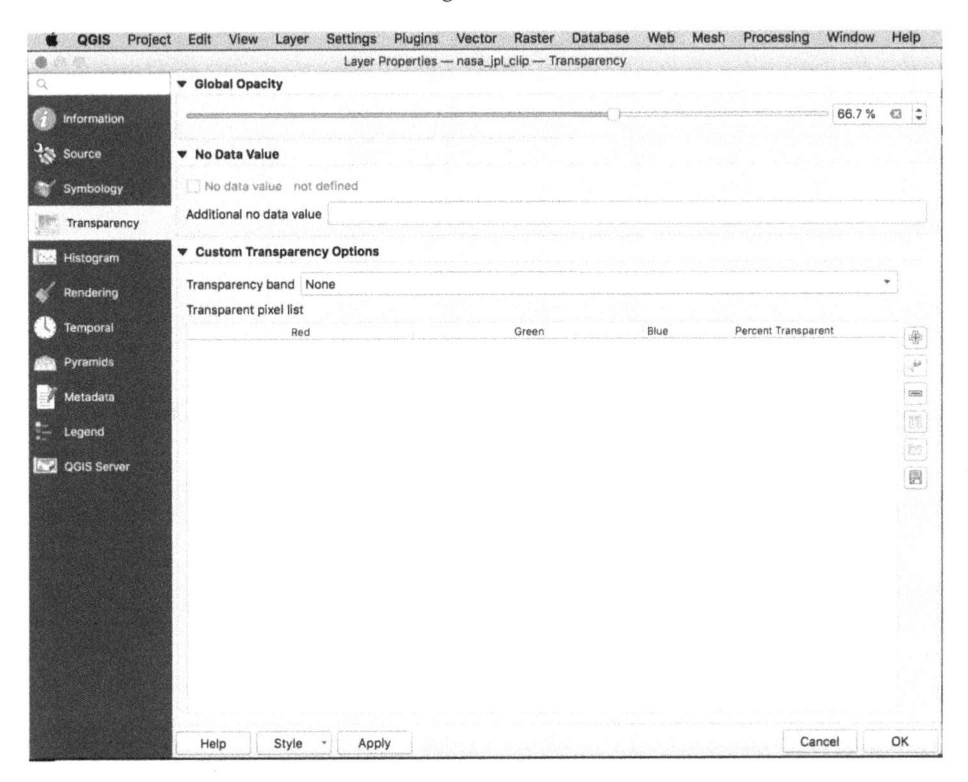

If you click *Apply*, it leaves you in the *Transparency* dialog and you can alter it again until you get what you want. If you click *OK* the dialog is closed. This is a common feature in many of the dialog boxes in QGIS.

> An even faster way to interactively style your layers is using the *Layer Styling* panel. Open it using the first icon at the top of the *Layers* panel. Using the styling panel you can set transparency, colors, labeling, and other styling properties. As you make changes, your map is instantly updated.

Now lets add more data. Go again to your `Cape GeoPackage` folder and highlight and drag several more of the files onto your map area. You can do this easily by clicking and dragging a group of files. Remember, if shapefiles, only get the .shp files, not the ones with other extensions such as .dbf and .shx.

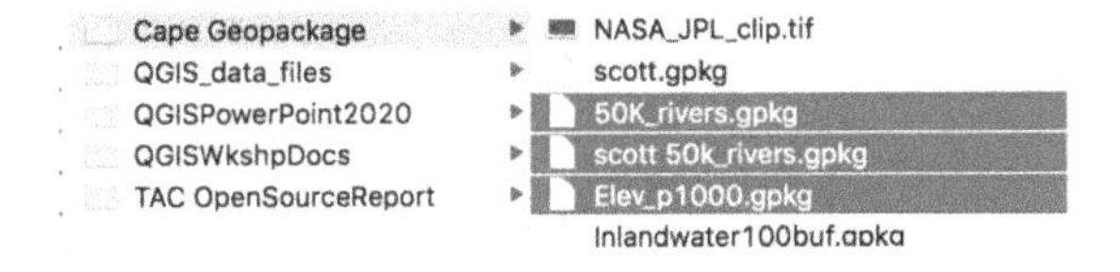

2.13 Zoom and Pan

You should now have all your vector files, as well as a partially transparent topo map over the satellite image. Zoom in and explore your data using the zoom and pan icons:

These are, left to right: Pan Map, Pan Map to Selection, Zoom In, Zoom Out, Zoom Full, Zoom to Selection, Zoom to Layer, Zoom to Native Resolution (of raster data only), Zoom Last, and Zoom Next. Note that you can do the same using the `View` menu, and many actions can be done using a combination of `Control (Ctrl)`, `Option (Alt)`, and other keys (hover your mouse over an icon to view function).

2.14 Saving your project

Now that you have populated a new QGIS project, it is very important to save it as a new QGIS project so you can come back to the same collection of data layers as a group. Go to `Project->Save` and type in `CapeGeoFinal` (or whatever you wish), pick where you want it to be saved, and hit return or click on `Save`.

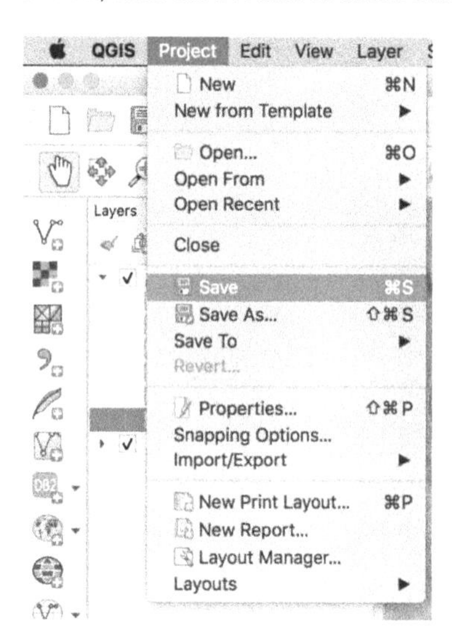

Remember where you saved it and use mnemonic names that actually represent what it is (don't call things kz99). You have now created a new QGIS project, and you can open this any time just as you last left it. The file will be `CapeGeoFinal.qgz` (or what you called it .qgz). The .qgz format saves your project as a zipped file.

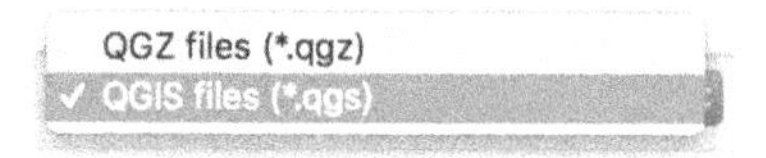

If you leave QGIS and want to start working with the same data and same zoom, etc., just navigate to that .qgs file, click on it, and it will then open QGIS, and open the project just as you left it, with exactly those layers, colors, level of zoom, etc. QGIS saves projects in a .qgs file, just as ArcGIS uses an .mxd file. You can also open it by opening QGIS and clicking on the `Projects->Open Recent` in the menu.

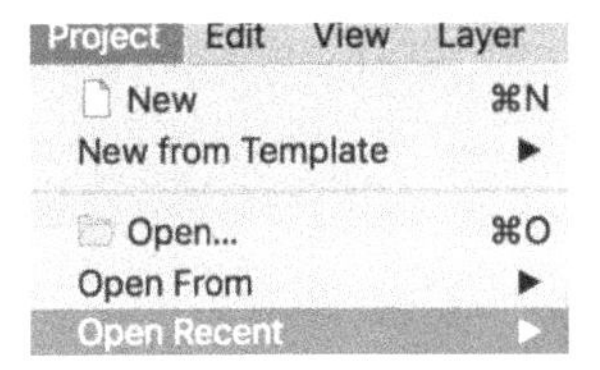

Be sure to save your project work often, as QGIS does not have an autosave feature. there is an AutoSaver plugin we will cover later. Once you have saved your project, go to the `Project->Project Properties` menu. Each QGIS project has properties, just as individual files do. This will show you your properties information and options for this particular saved project. Explore these:

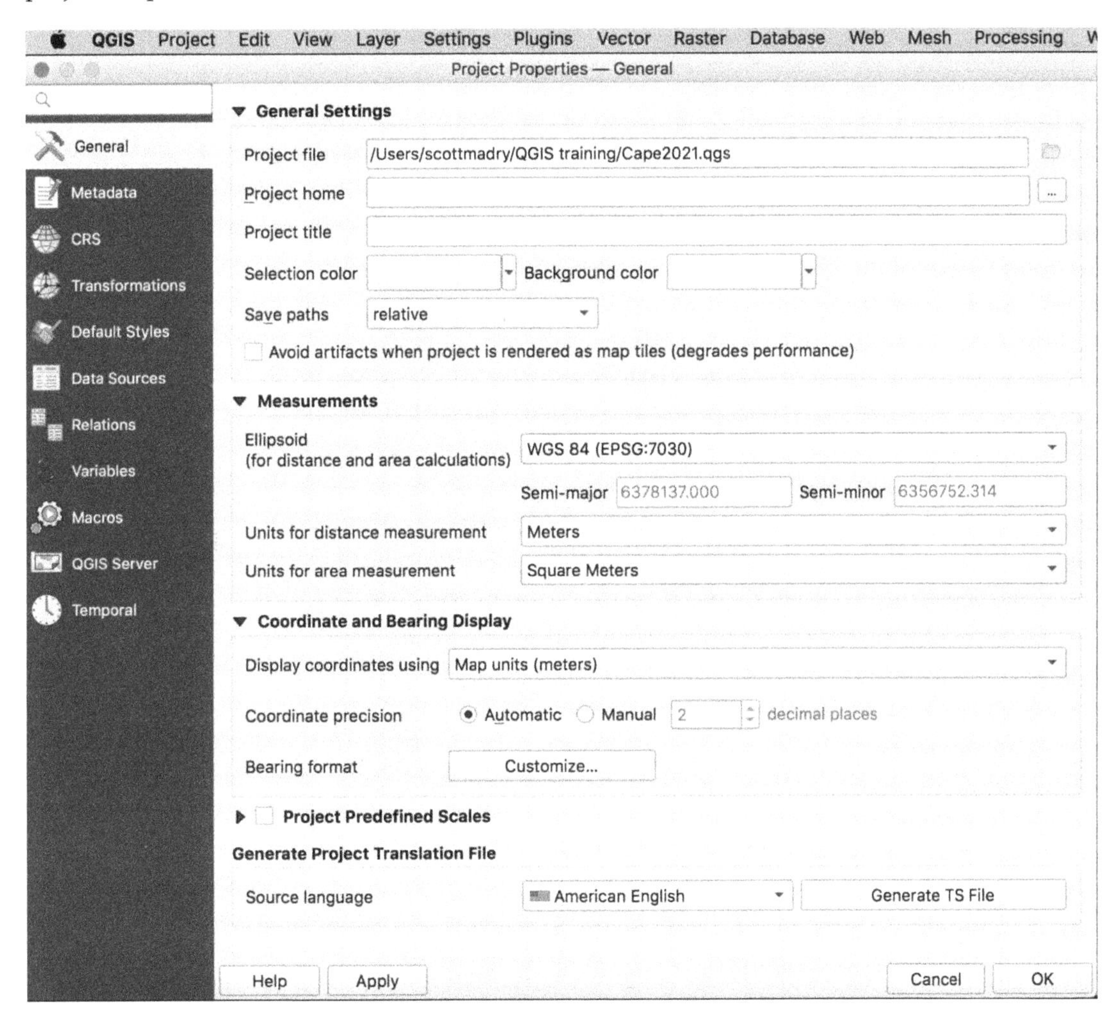

2.15 Coordinate Reference Systems (CRS)

This tab under *Project Properties* is where you can choose which units to use for distance and area measurements. Click on the *CRS* tab, and you will see that the Coordinate Reference System (CRS) for this project (based on the files we added), is WGS84 / UTM zone 35S. The UTM coordinate system is a commonly used system, and is appropriate for many GIS purposes. To learn more about it, see: `https://en.wikipedia.org/wiki/Universal_Transverse_Mercator_coordinate_system`

There are hundreds of CRS systems available in QGIS:

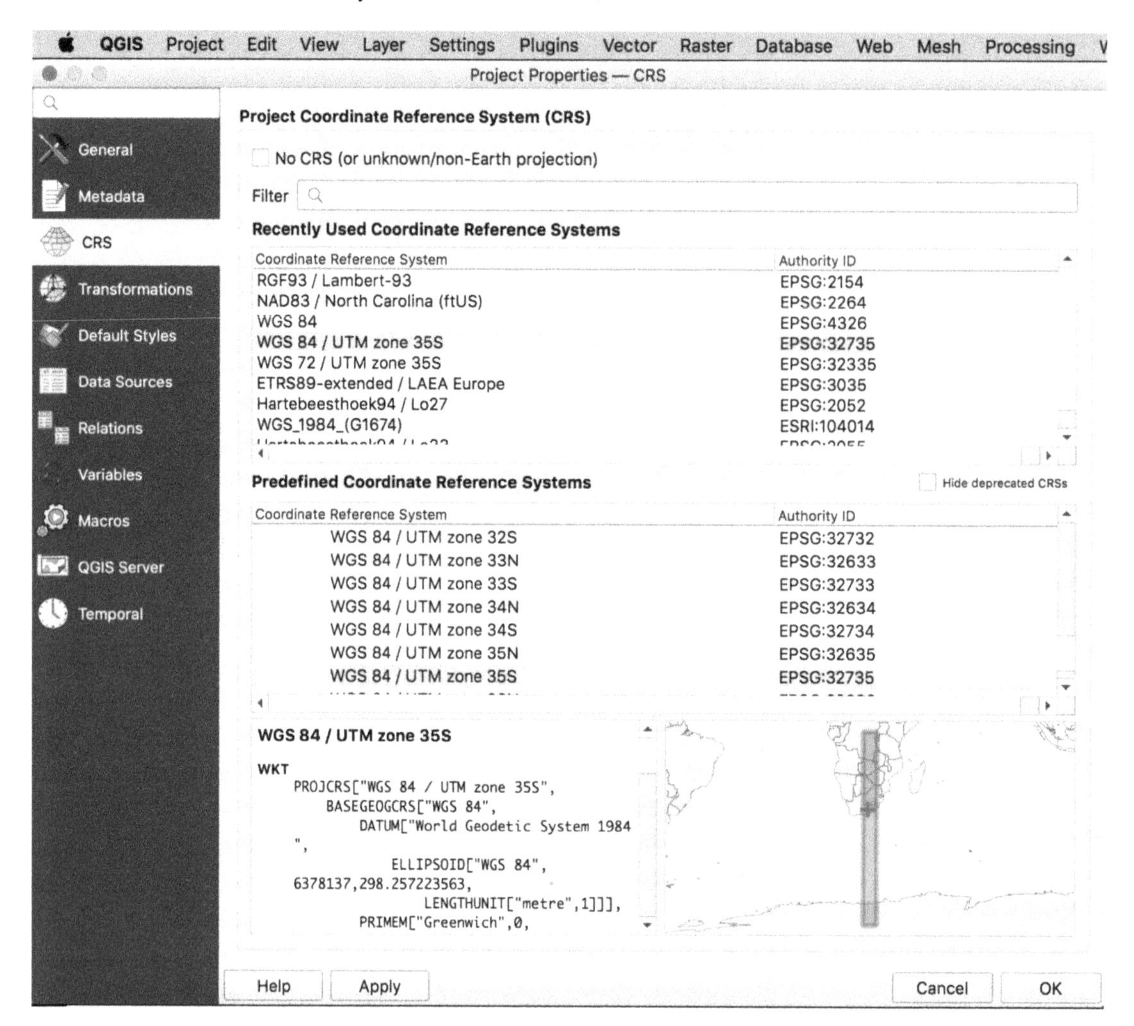

You can also create your own CRS. If you want to change this project's UTM projection, it is simple. Go to `Project->Properties->CRS`, and in the top line for *Filter*, type: *EPSG 2055* or *Hartesbeesthoek94*, which is the original South African projected datum.

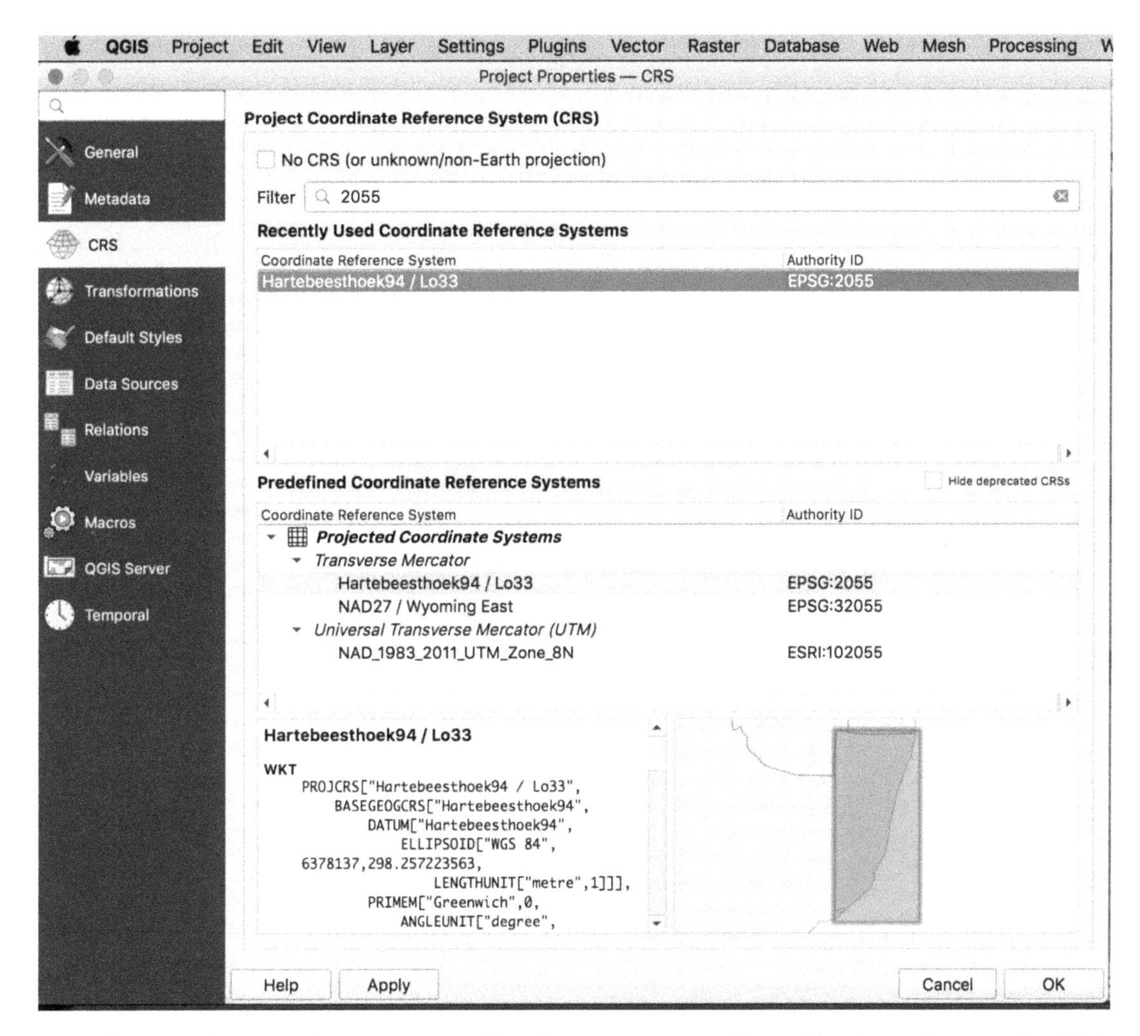

Note the map showing the area covered by the projection, indicated by the red rectangle. Click *OK* at the bottom and the project is now in that projection.

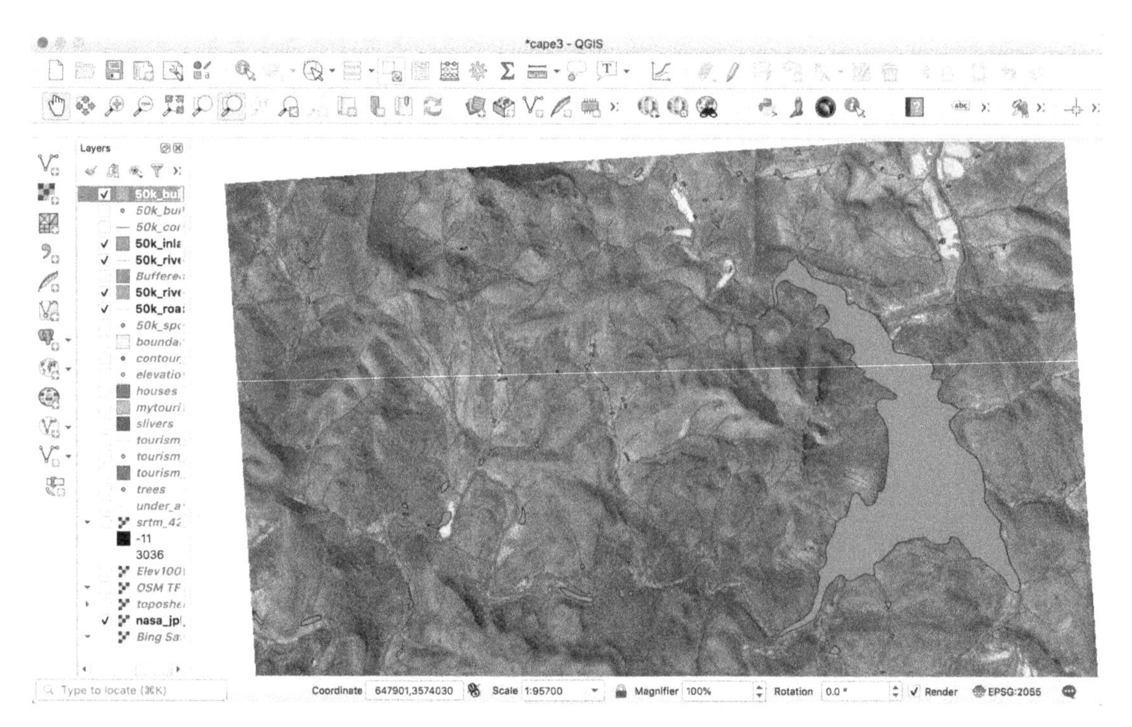

Note that the map is upside down because of the original, South African, national "south is up" map projection used in all early cadastral mapping. But this is strange to us northerners, so lets change it back to a UTM (Universal Transverse Mercator) metric projection, with north up. Go back to the *Project Properties* tab, and enter UTM zone 35S (EPSG 32335) in the filter line at top:

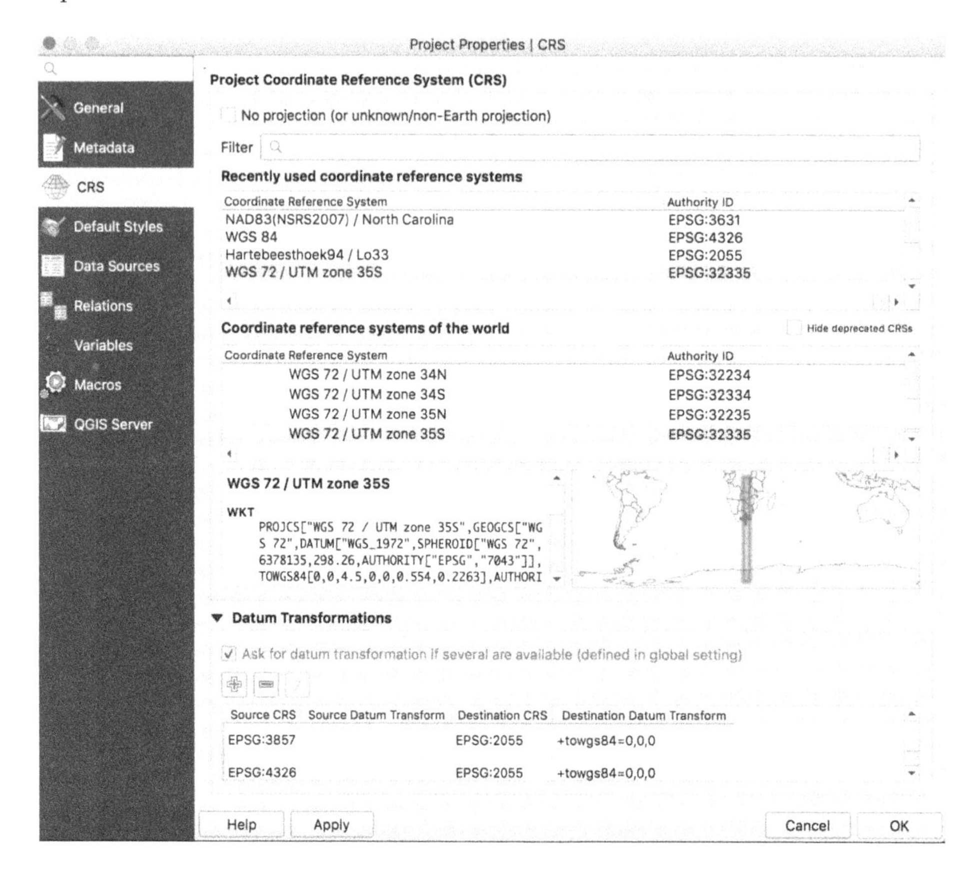

and click *OK*. Now north is up in our project and it's in a metric projection. You see that it shows you a map of where this projection is located, and that it covers our area in South Africa.

EPSG Codes

EPSG codes represent the EPSG (European Petroleum Study Group) Geodetic Parameter Dataset, or EPSG registry. This is a public registry of coordinate reference systems, ellipsoids of the Earth, and spatial transformations. Map projections and ellipsoids are assigned a unique code, which most GIS and remote sensing software systems can interpret and use. The EPSG codes were originally developed by the European Petroleum Standards Group in 1985. See (`https://en.wikipedia.org/wiki/EPSG_Geodetic_Parameter_Dataset` and `https://epsg.org`) for more information.

Note that you no longer need to enable 'on the fly' CRS transformations, as in previous QGIS versions. This is now always enabled, and it will automatically let you mix data with different CRS and view them together in one display, the current projection of your project. You can have data or images in different projections and they will all line up properly.

2.16 Project Properties

Click on the *General* tab and you can set the units, give your project a title, etc. In the *Data Sources* tab you can see the raster and vector layers included in your project:

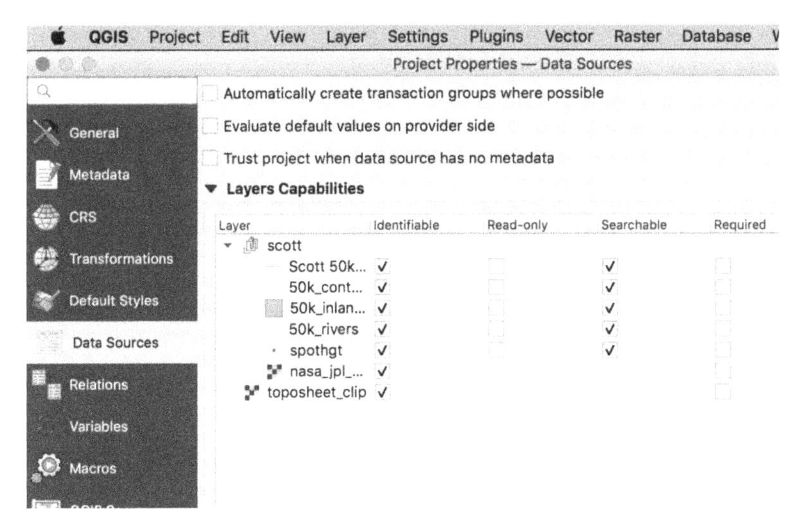

You can make files read only here, if you want to protect them. Note that you also have the option to create project metadata, which you should:

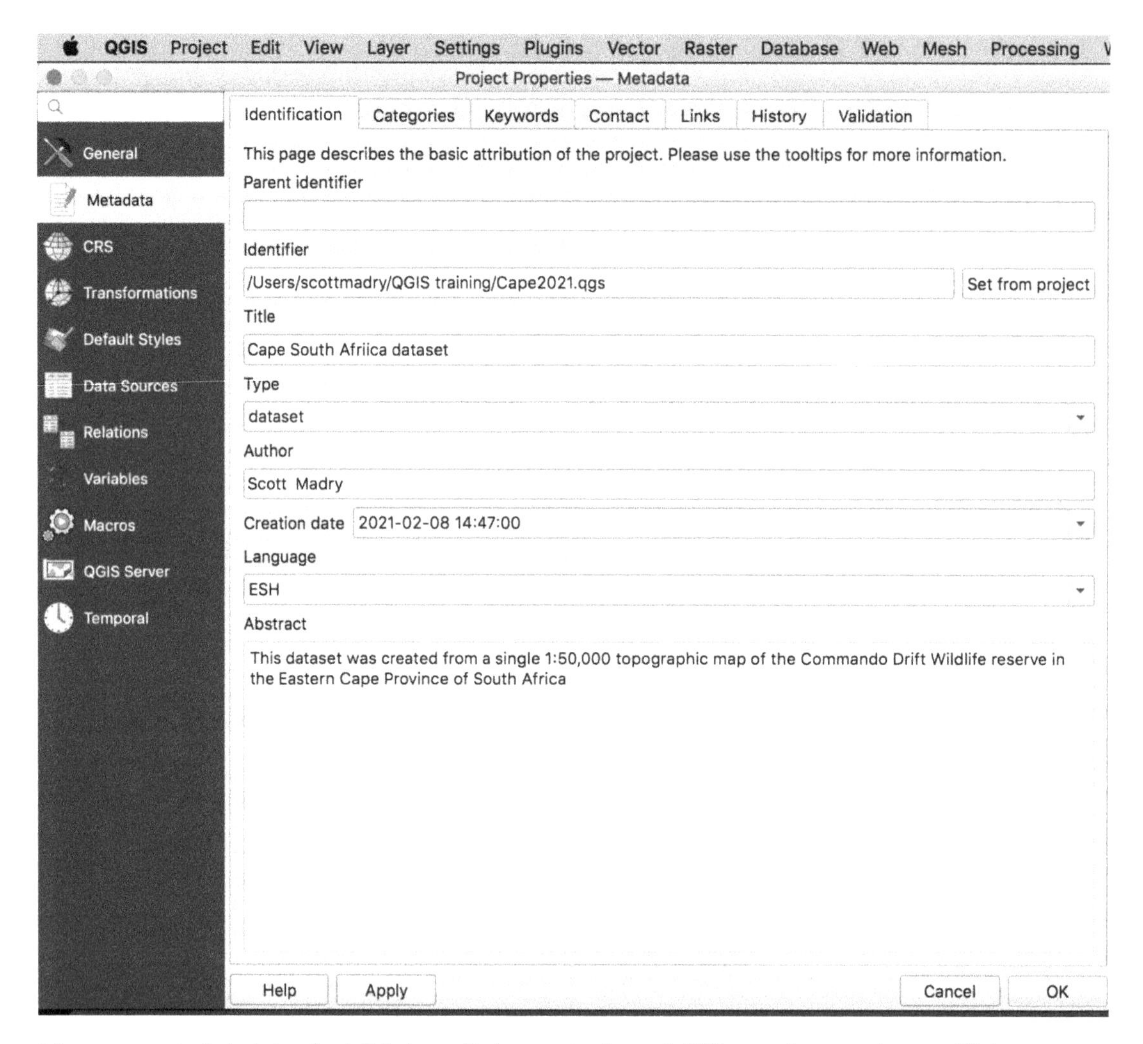

More on metadata later, but this is a vital aspect of good GIS practice, so always fill it out.

2.17 More Map Interaction Tools

Zoom in using the zoom tool and then roam around with the pan tool and explore the project area. Click on and off the topo map and image files, and some of the vectors. You can zoom to individual features by clicking on a layer name in the *Layers Panel* (try `50K_inland_waters`) so that it is highlighted, then click on the `Zoom to Extent` to see the full area. Clicking on the *Identify Features* tool and then clicking on the blue lake will display the attributes for the feature like this:

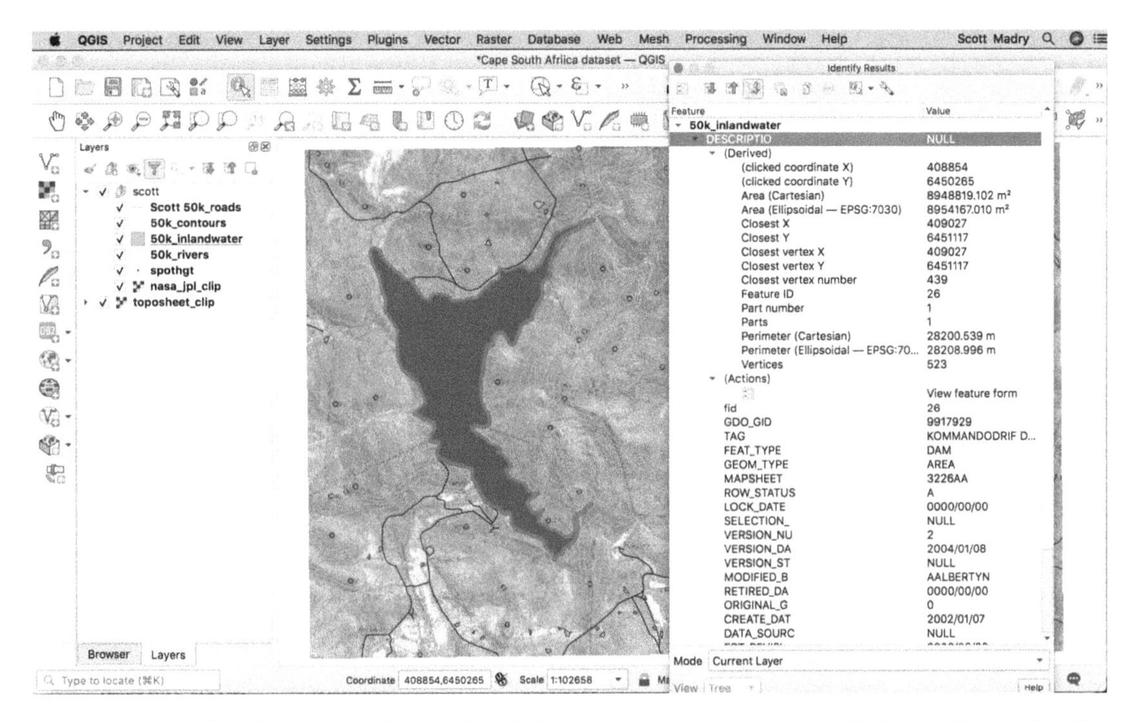

The feature will be highlighted in red and the pop-up window with all the attributes for the layer will be displayed. Note there are several options available in the *Identify Results* tab. Hover over the icons to see what they do. You can remove this pop-up and the red outline of the lake by clicking on the yellow and red *Clear Results* icon, as shown below:

2.18 Vector Attributes and Attribute Tables

Attributes are attached to features (points, lines, and polygons) in your vector GIS files. QGIS uses the GeoPackage backed by a SQLite database, but also supports others, including PostgresSQL, Oracle, DB2, and MSSQL. The attributes are the non-spatial content of the database. Double-click on the `50k_contours` layer in the *Layers* panel to open the *Layer Properties* dialog and click on *Symbology* to change the color and line style of the layer.

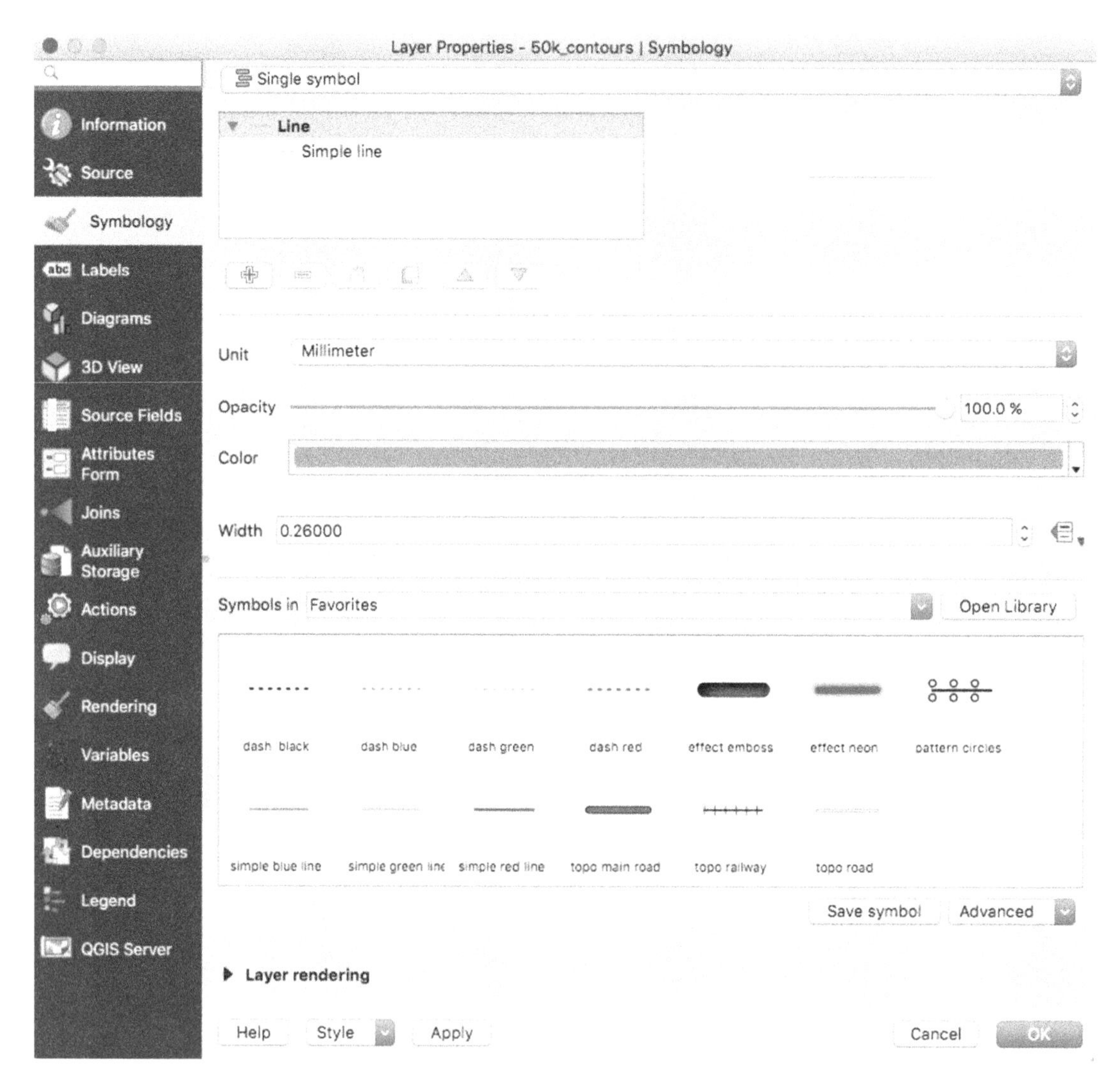

You can see the layer is currently rendered in brown and is hard to see. Click on the down arrow at right in the *Color* tab, and change the color to black, then click *OK*. You should be able to see the contours better now. Note that there are several vector line options to choose from as well.

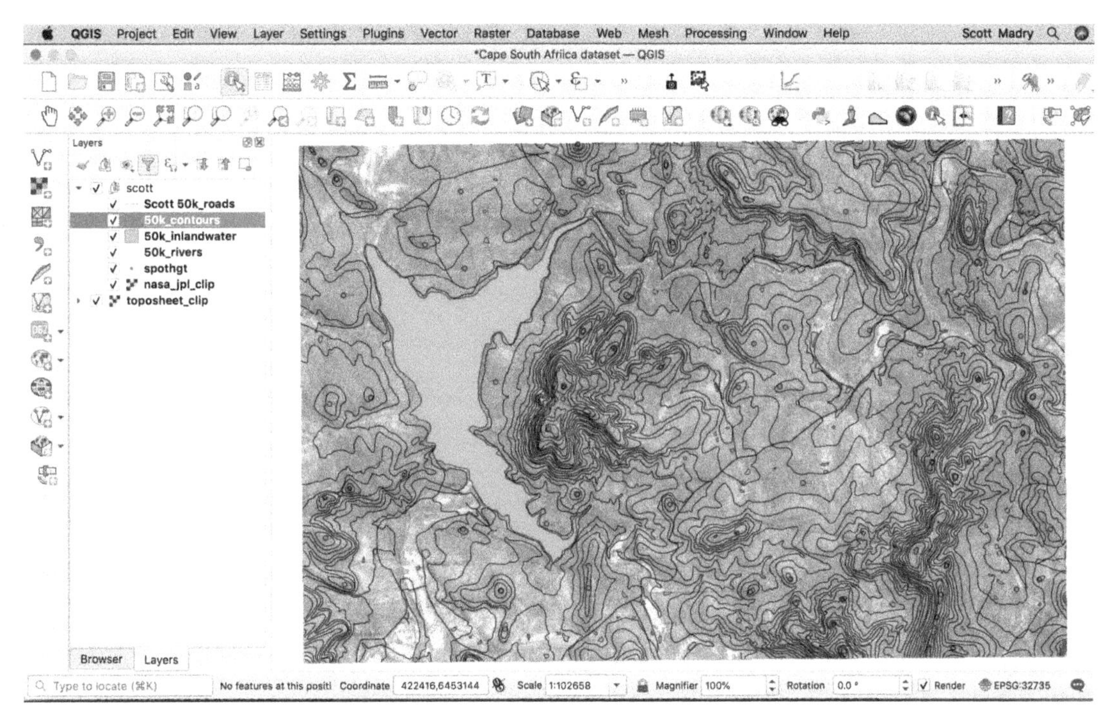

Our `50k_contours` layer contains elevation contours, extracted from the 1:50,000 topo map. You can view and edit the attributes of a vector layer by right-clicking on it in the *Layers* panel and choosing `Open Attribute Table`. You can now see the attributes for the layer:

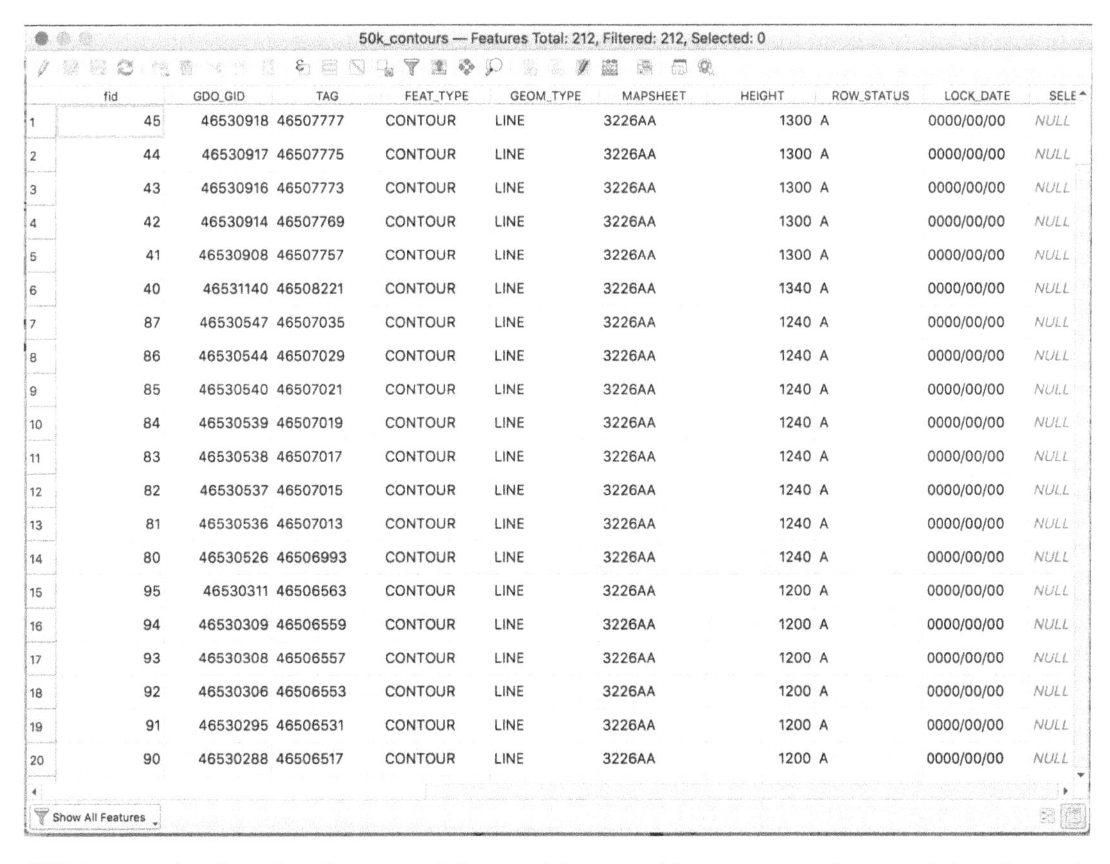
50k_contours — Features Total: 212, Filtered: 212, Selected: 0

	fid	GDO_GID	TAG	FEAT_TYPE	GEOM_TYPE	MAPSHEET	HEIGHT	ROW_STATUS	LOCK_DATE	SELE
1	45	46530918	46507777	CONTOUR	LINE	3226AA	1300	A	0000/00/00	NULL
2	44	46530917	46507775	CONTOUR	LINE	3226AA	1300	A	0000/00/00	NULL
3	43	46530916	46507773	CONTOUR	LINE	3226AA	1300	A	0000/00/00	NULL
4	42	46530914	46507769	CONTOUR	LINE	3226AA	1300	A	0000/00/00	NULL
5	41	46530908	46507757	CONTOUR	LINE	3226AA	1300	A	0000/00/00	NULL
6	40	46531140	46508221	CONTOUR	LINE	3226AA	1340	A	0000/00/00	NULL
7	87	46530547	46507035	CONTOUR	LINE	3226AA	1240	A	0000/00/00	NULL
8	86	46530544	46507029	CONTOUR	LINE	3226AA	1240	A	0000/00/00	NULL
9	85	46530540	46507021	CONTOUR	LINE	3226AA	1240	A	0000/00/00	NULL
10	84	46530539	46507019	CONTOUR	LINE	3226AA	1240	A	0000/00/00	NULL
11	83	46530538	46507017	CONTOUR	LINE	3226AA	1240	A	0000/00/00	NULL
12	82	46530537	46507015	CONTOUR	LINE	3226AA	1240	A	0000/00/00	NULL
13	81	46530536	46507013	CONTOUR	LINE	3226AA	1240	A	0000/00/00	NULL
14	80	46530526	46506993	CONTOUR	LINE	3226AA	1240	A	0000/00/00	NULL
15	95	46530311	46506563	CONTOUR	LINE	3226AA	1200	A	0000/00/00	NULL
16	94	46530309	46506559	CONTOUR	LINE	3226AA	1200	A	0000/00/00	NULL
17	93	46530308	46506557	CONTOUR	LINE	3226AA	1200	A	0000/00/00	NULL
18	92	46530306	46506553	CONTOUR	LINE	3226AA	1200	A	0000/00/00	NULL
19	91	46530295	46506531	CONTOUR	LINE	3226AA	1200	A	0000/00/00	NULL
20	90	46530288	46506517	CONTOUR	LINE	3226AA	1200	A	0000/00/00	NULL

Show All Features

GIS data can be thought of as a spatial spreadsheet, and here you see the actual data for each component of the layer. You also see a new set of icons along the top of the attribute window.

Hover over the icons to see their functions. These allow you to edit attributes, sort, delete, or add columns, etc. Be careful—doing so will alter your data. Always work with a copy, and keep your original data safe so that it can be reloaded if you make a mistake. To begin editing the attribute table, click on the yellow pen at left. One of the benefits of using QGIS is that all your basic functions are accessed directly in this interface, you do not have to go back and forth between different interfaces like in ArcGIS with the standard Arc interface, Arc Catalog, and Arc Toolbox.

You can zoom to a given feature by clicking the row first (the number at far left), and then clicking on the magnifying glass icon at top center left. Click on the first line in the attribute table, by clicking on the number 6 at left and the entire line will be highlighted in blue:

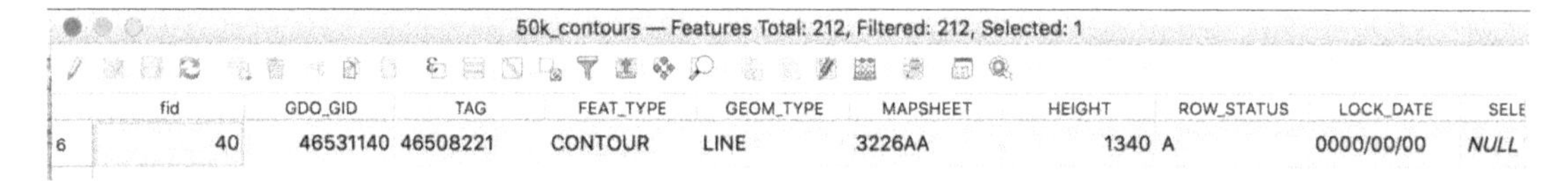

Next, click on the magnifying glass icon . You will see this:

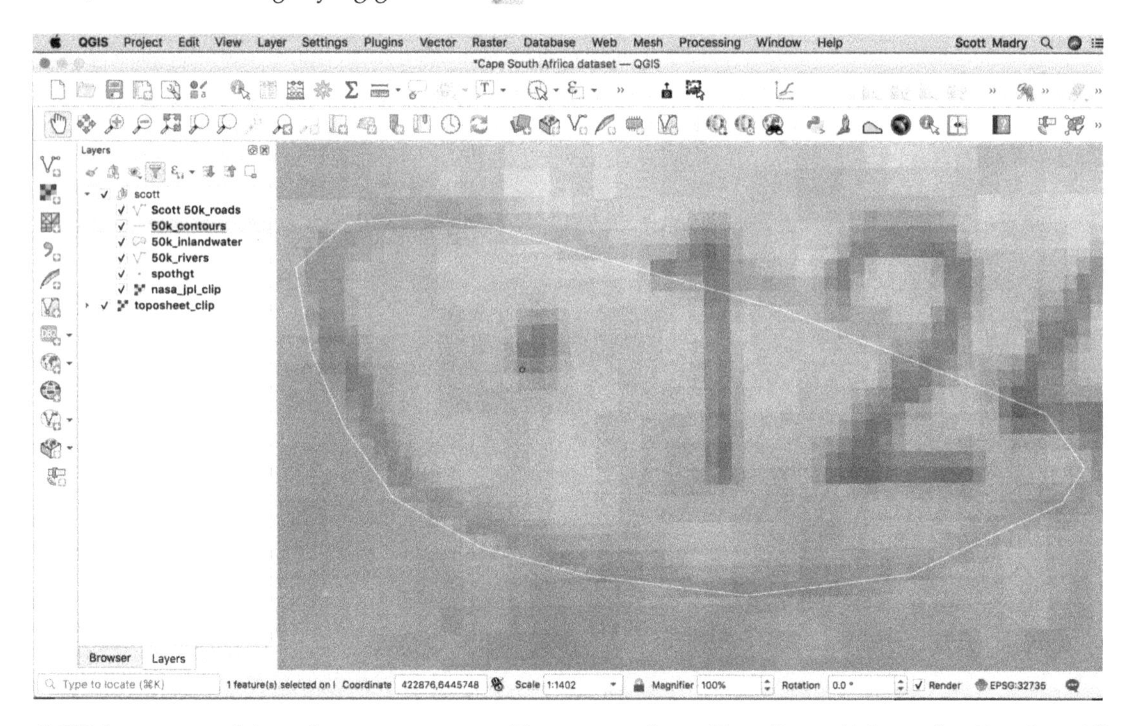

QGIS has zoomed into that one contour. To remove the yellow line, click on the *Deselect All Features* icon as shown below:

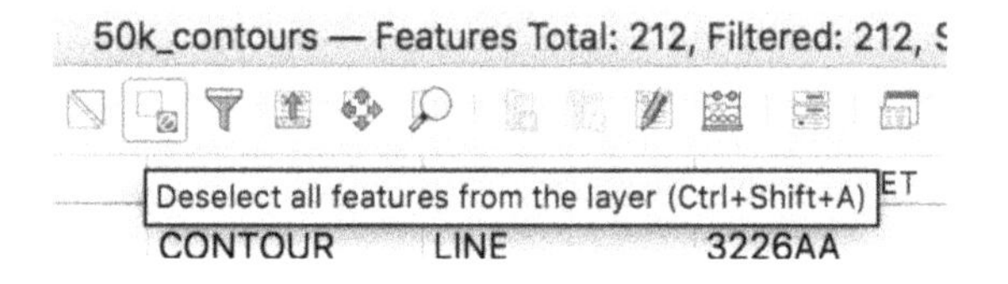

Click on the *Zoom Out* icon to see a slightly larger view. You can edit attributes, search, sort, add new columns, etc. This is one of the most powerful aspects of GIS.

2.19 The Identify Tool

You can also "drill down" into your attribute data from the map interface. Right-click on the contours layer and click on the first option, `Zoom to Layer`, to see the entire study area, then click on the *Identify Features* tool. Click on any individual contour to see the attributes:

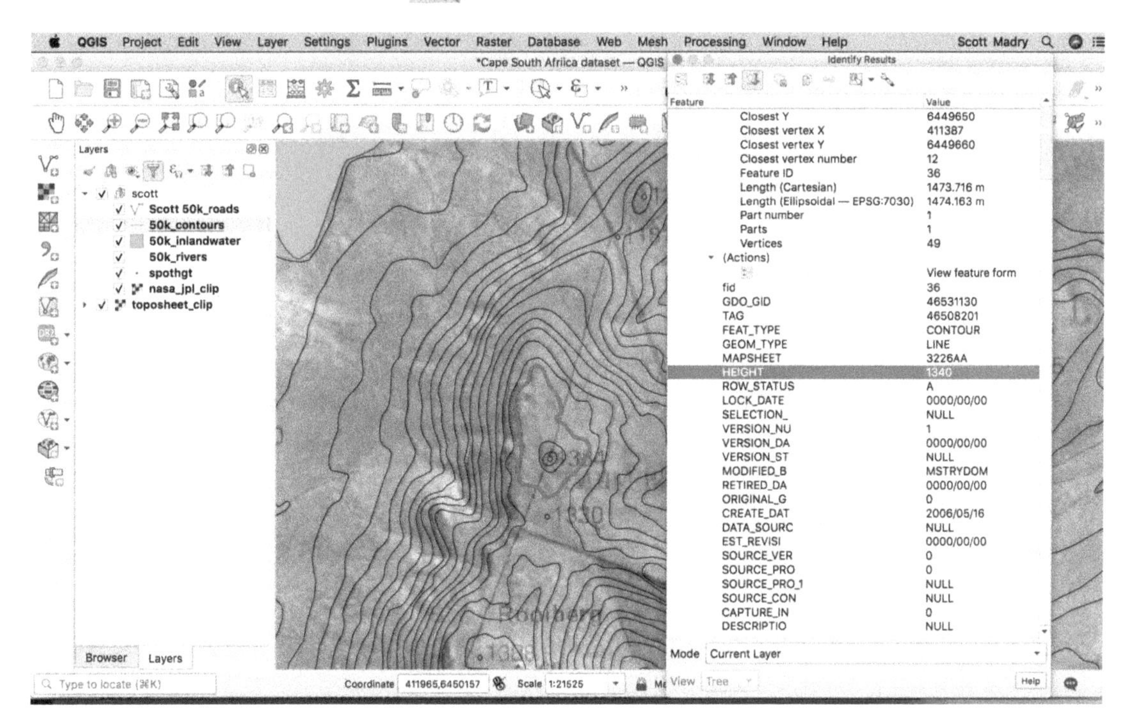

The individual contour is highlighted and the pop-up window displays the attribute data. Here, we are looking at a contour with the height of 1340 meters above sea level. To return to the QGIS interface, click on the yellow and red *Clear Results* icon at the top center of the pop-up, then close the *Identify Results* window.

You can also open the attribute table by clicking on the layer in the *Layers* panel to highlight it, then click on the attribute icon.

2.20 The Screen Measurement Icon

Next, let us look as some additional icons at the top. You can do simple, on-screen measurements using the 'measure' icon. Click on the down-arrow next to the icon and you can set it to measure line lengths, areas, or angles. A pop-up appears and you can measure one or multiple segments:

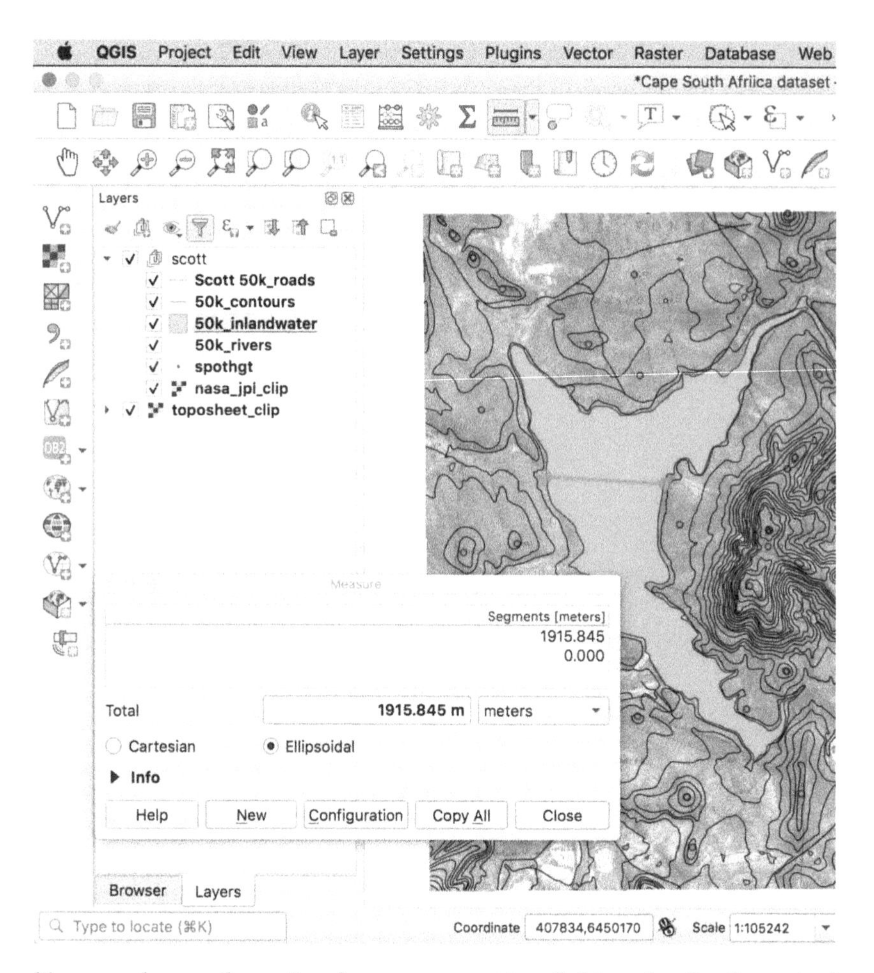

You can change the units of measurement by clicking the *Configuration* button, which will open the `Settings->Options->Map Tools` dialog.

2.21 Select Features

There is a useful *Select Features* tool [icon] that allows you to select either single or multiple features in a vector file in several ways: individual feature()s, within a polygon, by radius, or freehand.

Click on `50k_contours` in the *Layers* panel to highlight the layer, then click on the *Select Features* icon and choose `Select Features by Radius`. Click on the map and draw the radius with the mouse. The contours that are within the area will be highlighted:

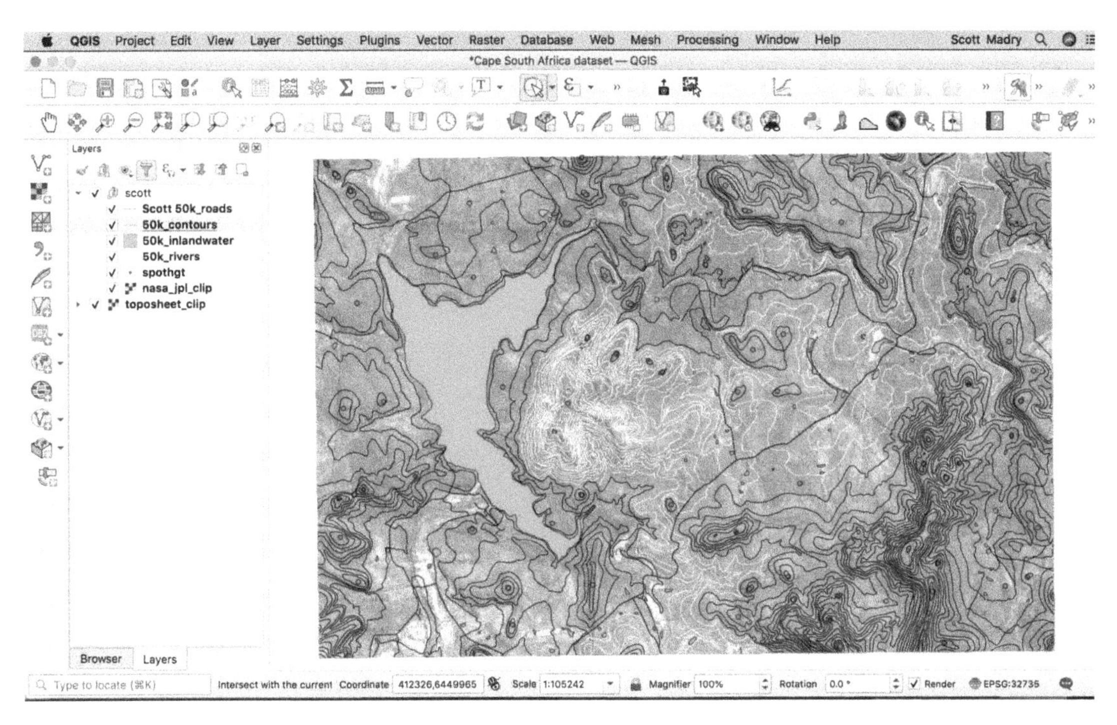

Right-click on `50k_contours` in the *Layers* panel, choose `Open Attribute Table`, and the selected contours will be highlighted in the attribute table.

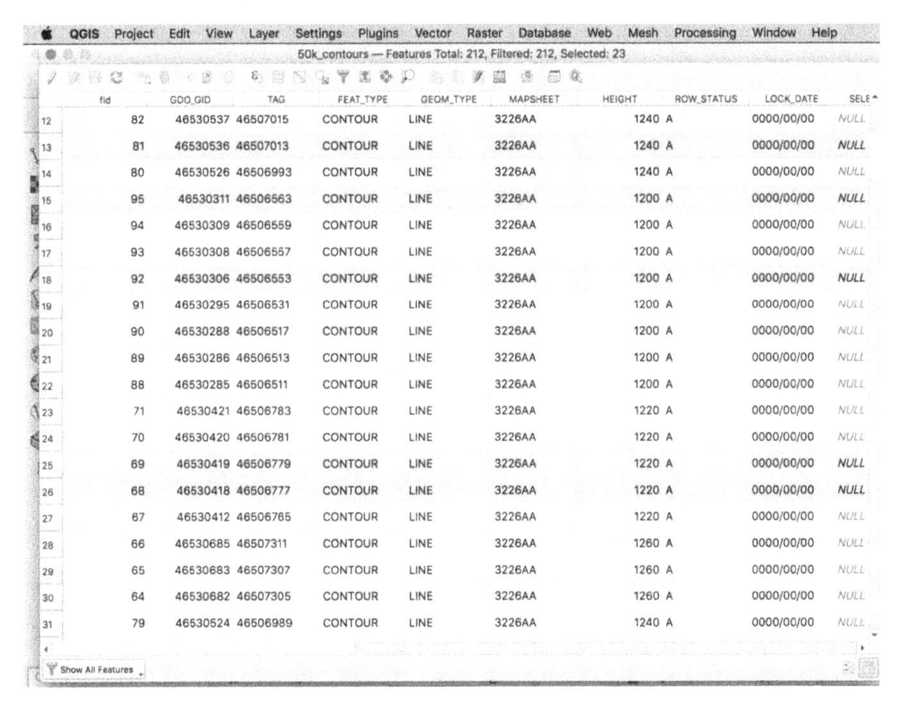

	fid	GDO_GID	TAG	FEAT_TYPE	GEOM_TYPE	MAPSHEET	HEIGHT	ROW_STATUS	LOCK_DATE	SELE
12	82	46530537	46507015	CONTOUR	LINE	3226AA	1240	A	0000/00/00	NULL
13	81	46530536	46507013	CONTOUR	LINE	3226AA	1240	A	0000/00/00	NULL
14	80	46530526	46506993	CONTOUR	LINE	3226AA	1240	A	0000/00/00	NULL
15	95	46530311	46506563	CONTOUR	LINE	3226AA	1200	A	0000/00/00	NULL
16	94	46530309	46506559	CONTOUR	LINE	3226AA	1200	A	0000/00/00	NULL
17	93	46530308	46506557	CONTOUR	LINE	3226AA	1200	A	0000/00/00	NULL
18	92	46530306	46506553	CONTOUR	LINE	3226AA	1200	A	0000/00/00	NULL
19	91	46530295	46506531	CONTOUR	LINE	3226AA	1200	A	0000/00/00	NULL
20	90	46530288	46506517	CONTOUR	LINE	3226AA	1200	A	0000/00/00	NULL
21	89	46530286	46506513	CONTOUR	LINE	3226AA	1200	A	0000/00/00	NULL
22	88	46530285	46506511	CONTOUR	LINE	3226AA	1200	A	0000/00/00	NULL
23	71	46530421	46506783	CONTOUR	LINE	3226AA	1220	A	0000/00/00	NULL
24	70	46530420	46506781	CONTOUR	LINE	3226AA	1220	A	0000/00/00	NULL
25	69	46530419	46506779	CONTOUR	LINE	3226AA	1220	A	0000/00/00	NULL
26	68	46530418	46506777	CONTOUR	LINE	3226AA	1220	A	0000/00/00	NULL
27	67	46530412	46506765	CONTOUR	LINE	3226AA	1220	A	0000/00/00	NULL
28	66	46530685	46507311	CONTOUR	LINE	3226AA	1260	A	0000/00/00	NULL
29	65	46530683	46507307	CONTOUR	LINE	3226AA	1260	A	0000/00/00	NULL
30	64	46530682	46507305	CONTOUR	LINE	3226AA	1260	A	0000/00/00	NULL
31	79	46530524	46506989	CONTOUR	LINE	3226AA	1240	A	0000/00/00	NULL

To clear this selection click on the *Deselect all features...* icon in the attribute table toolbar or the *Deselect Features* icon on the QGIS interface.

2.22 Map Decorations

You can add a north arrow, text, grid, scale bar, and more to your display from the `View->Decorations` menu:

Be sure to click the *Enable Scale Bar* and *Enable North Arrow* checkbox on the individual dialogs to make them visible on the map. You can use the copyright tool for any text, not just a ©.

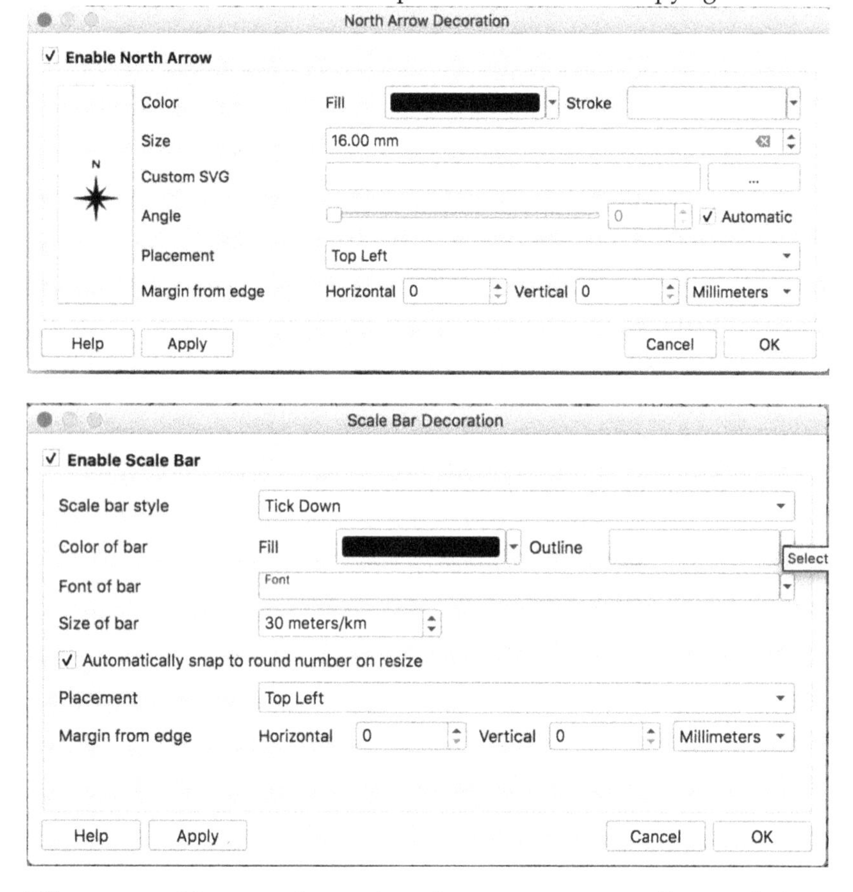

When you add a map decoration, they are automatically rescaled as you zoom and pan. In most cases, you'll want to add a north arrow and scale bar to your final map.

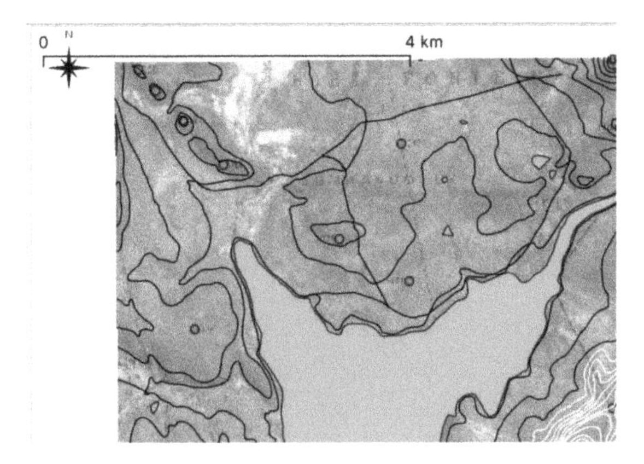

Under the `View` menu you will see you can do many of the map manipulation functions that are also available on the toolbar icons. You can remove the icons entirely from your display if you wish, by clicking on `View->Toolbars`:

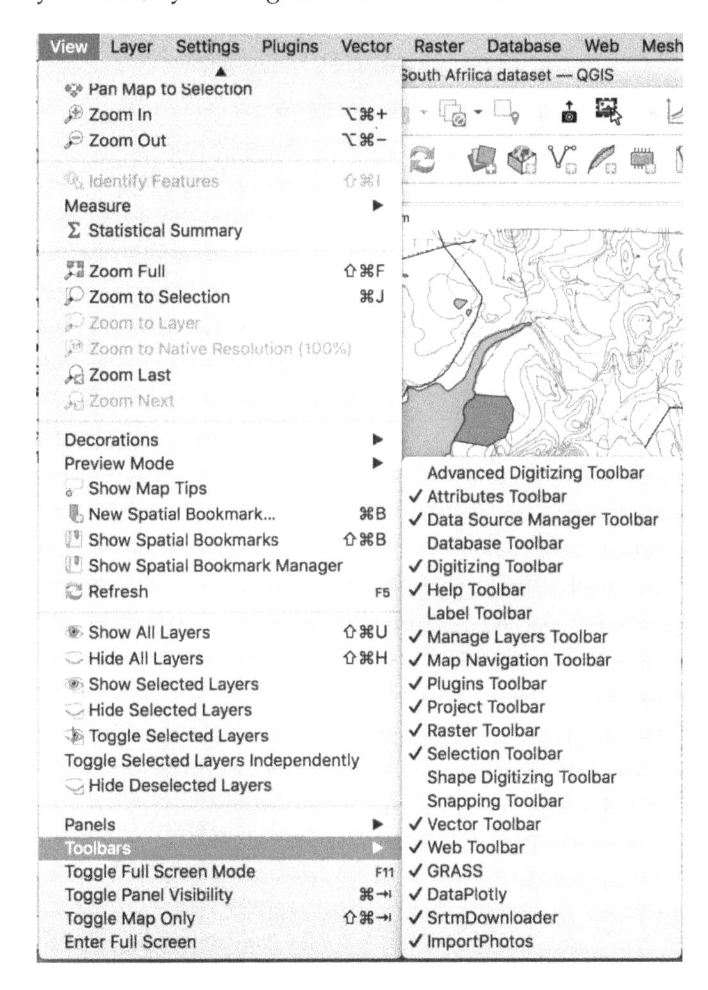

2.23 Spatial Bookmarks

QGIS has a very useful Spatial Bookmark feature that lets you create stored bookmarks so you can return to a given location and zoom level. Zoom into a small area and then click on the *New Bookmarks* icon, or from the menu, choose `View->New Spatial Bookmark`. Enter the name for the bookmark and adjust any other settings (extent, CRS, save location) you desire. Click *Save* to save the new bookmark and close the dialog.

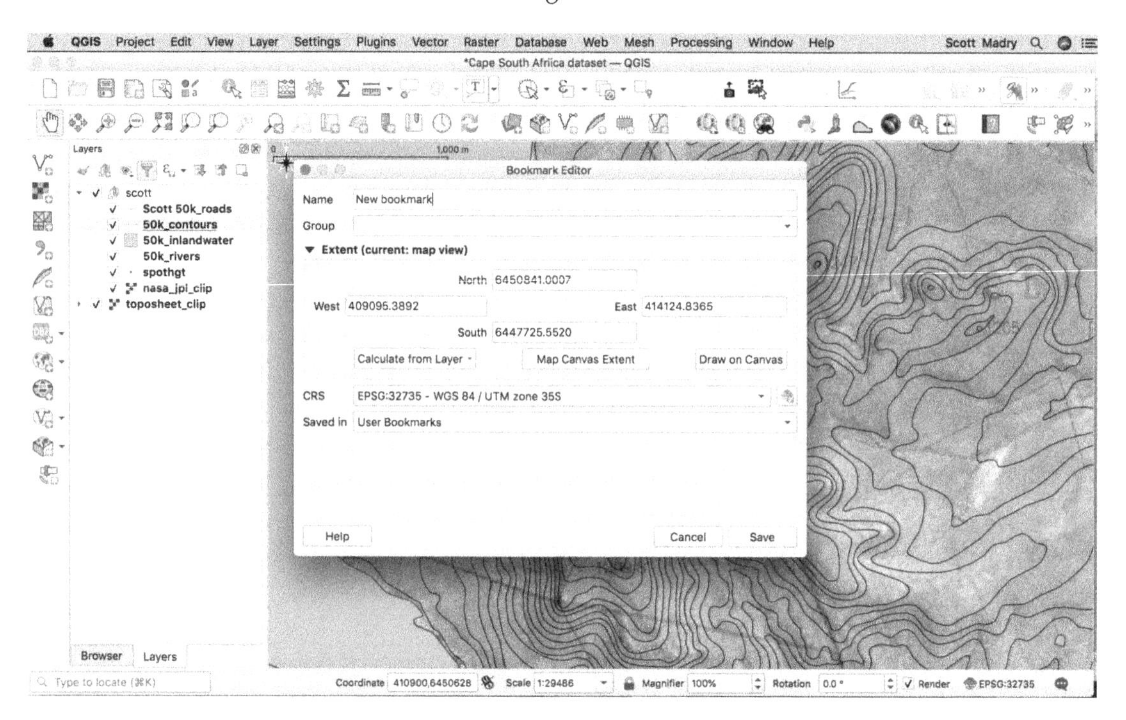

There is also a *Spatial Bookmarks Manager* panel—open it by choosing `View->Panels->Spatial Bookmarks Manager` from the menu. You can use the icons at the top of the bookmarks manager to add new bookmarks, delete, zoom to different ones, import/export, etc. To go to an existing bookmark, just click on it in the *Spatial Bookmarks Manager* and click the *Zoom to bookmark* (magnifying glass) icon.

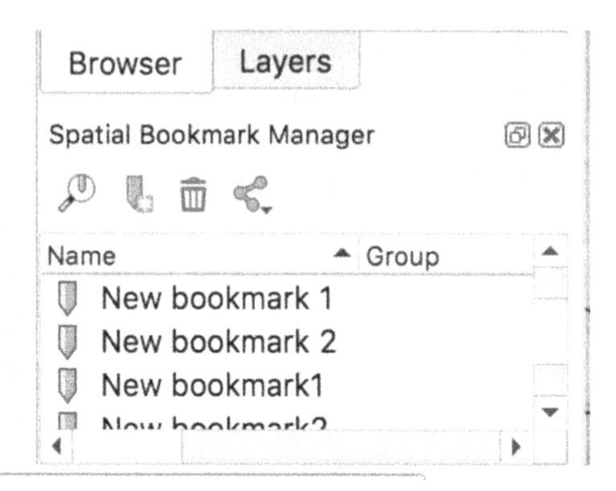

You can have an unlimited number of bookmarks, but be aware that more than one can have the same name. Also, note that your bookmarks now appear in your *Browser* Panel. Click on the *Show Spatial Bookmarks* icon to go to your bookmarks in the *Browser* panel. Double-click on the one you want to go to, or right-click on and choose `Zoom to Bookmark`.

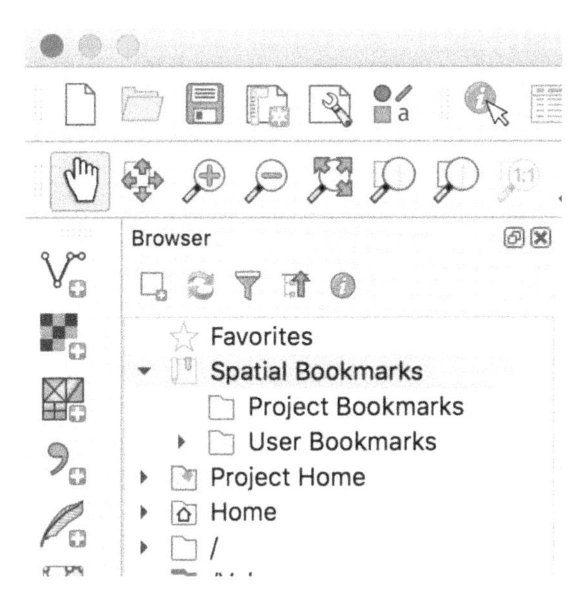

If you right-click on a bookmark here, you can also edit or delete it.

2.24 Text Annotation

QGIS has a simple and useful text annotation function. Click on the icon [T] and you will see the following menu:

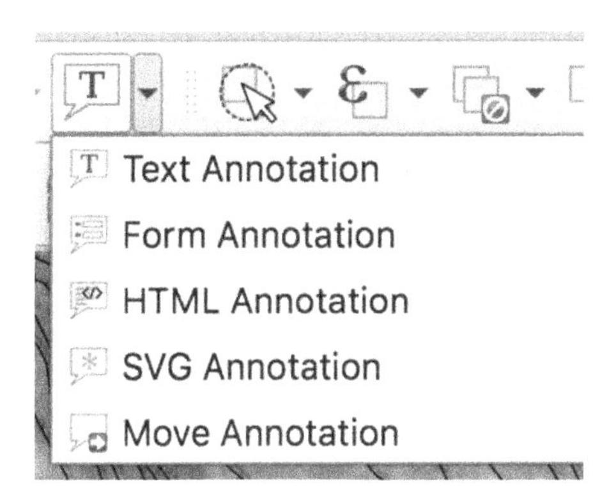

Lets put a balloon window in our map showing the lake. Click on `Text Annotation` [T] then click on the center of the lake and it will create a text box. Now double-click within the box and you will see the *Annotation Text* dialog:

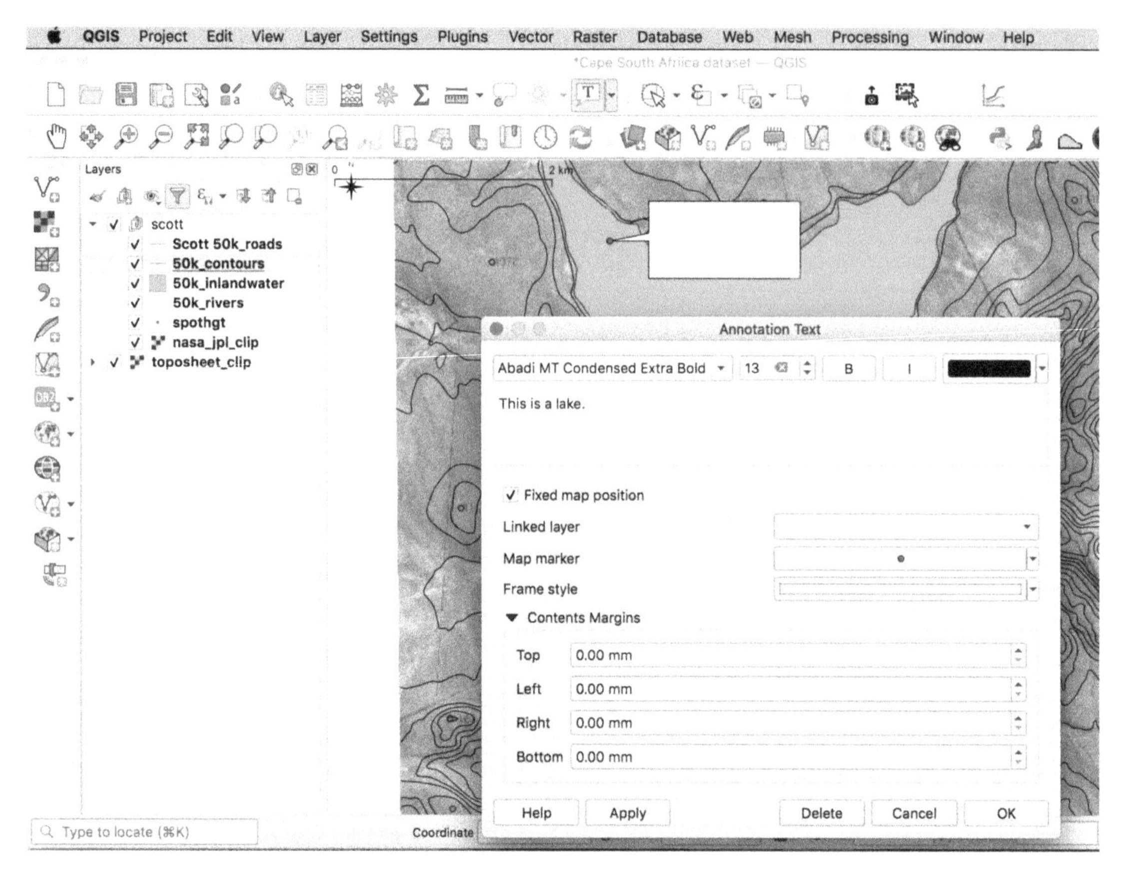

Enter the text you want, then set the font, color for the frame, etc. and click *OK*. The new text appears in the box on the map. Grab a corner and resize it. Click in the center and move it. To remove a text box, just double-click inside the box and click on *Delete*.

Click on the down arrow in the bottom right corner of the tool T and click on `Move Annotation`. You can now click on the red dot point and move it. There is also a `Form Annotation` option, where you can create your own annotation forms or link to a text file. This is useful to automatically annotate multiple features like city point files or contours. There are also SVG and HTML annotation options.

2.25 Actions

QGIS provides the ability to perform an external action based on the attributes of a feature. This can be used to perform any number of actions—for example, running a program with arguments built from the attributes of a feature or passing parameters to a web reporting tool. Actions are useful when you frequently want to run an external application or view a web page based on one or more values in your vector layer.

An example is performing a search based on an attribute value, or to open a website and run a Google or Wikipedia query on a name. To initiate an Action, right-click on a file in the *Layers* panel and highlight the *Actions* tab at left. Click on *Create Default Actions* and it will populate common actions such as opening a file, searching the internet based on an attribute, or starting a Python script:

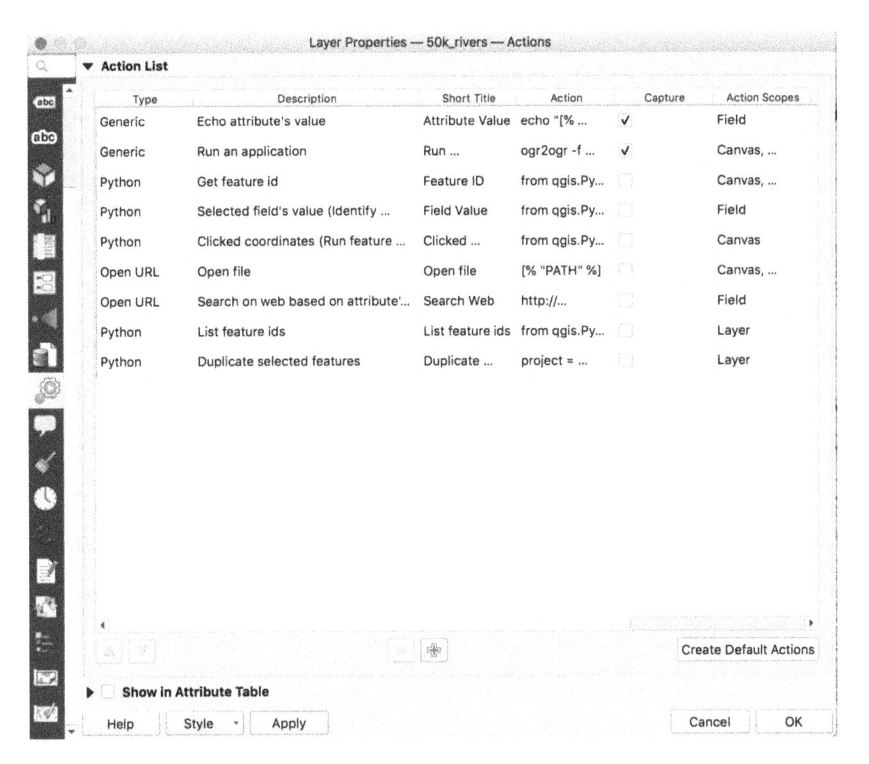

If you click on the green plus, a new dialog box to create an action will open:

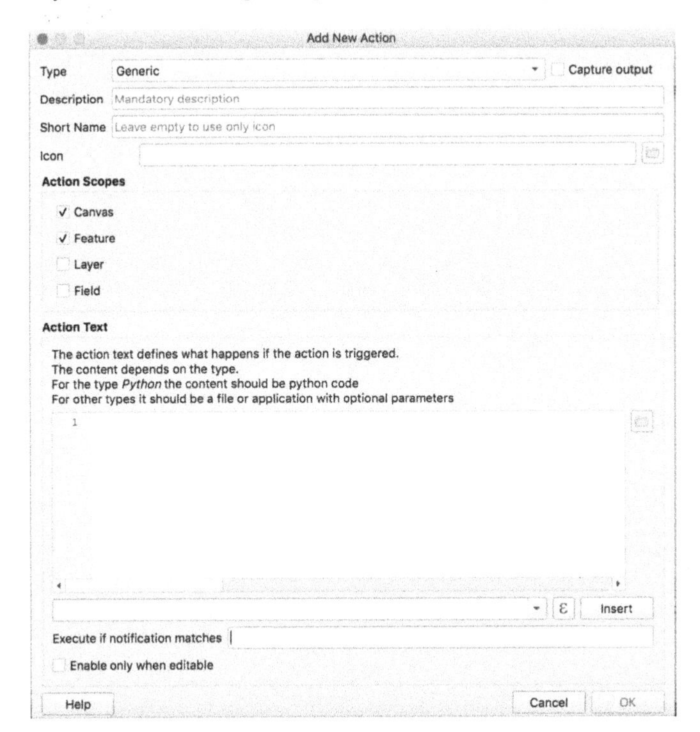

There is an actions icon, , with menu items for each action you have enabled. It is grayed out until you have enabled one or more actions.

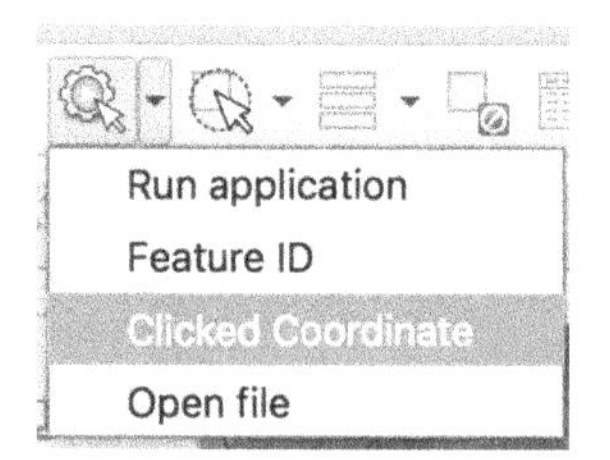

2.26 Labels

We can place labels on our map using attribute values. Right-click on the `50k_spothgt` layer and choose `Properties`, then click on the `Labels` tab:

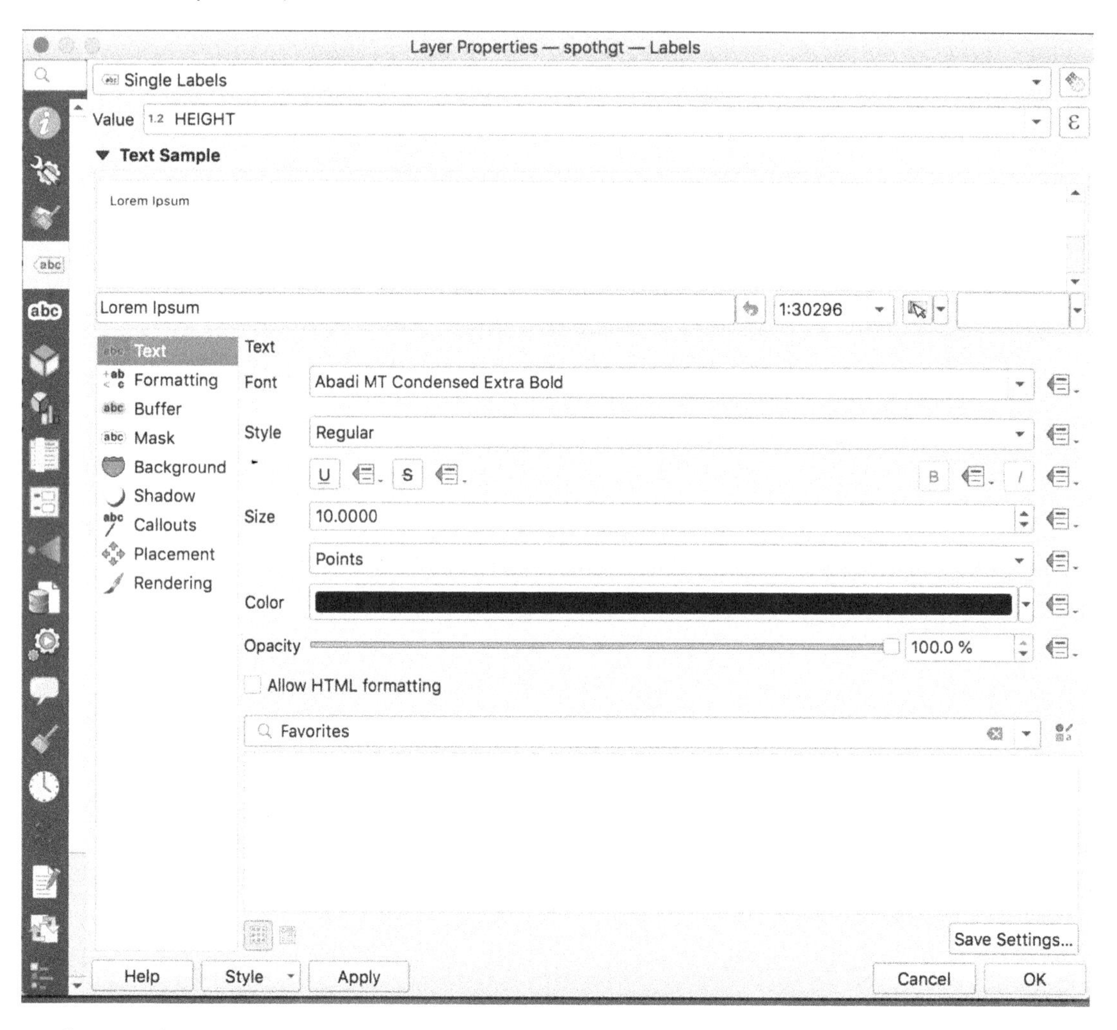

At the top, choose *Single Labels*, and next, choose the *Value height*, and click *OK*. Note that you can change the color, size, background, placement, etc.

2.27 Diagrams

QGIS lets us create pie charts and text diagrams based on the data in our layers and automatically display these. Right-click on the `50k_spothgt` layer and choose `Properties`, then click the *Diagrams* tab along the left side of the window.

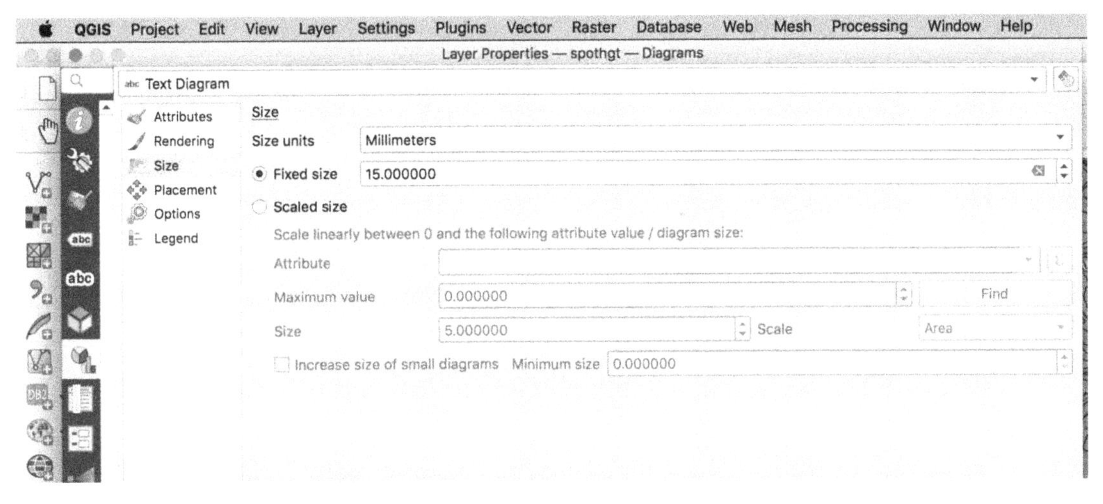

Choose *Text Diagram* from the drop-down box (if we had area data, we could pick *Pie Chart*). Click on *Attributes*, and click the attribute *HEIGHT* and click on the plus icon at right to add it to your list, as shown below:

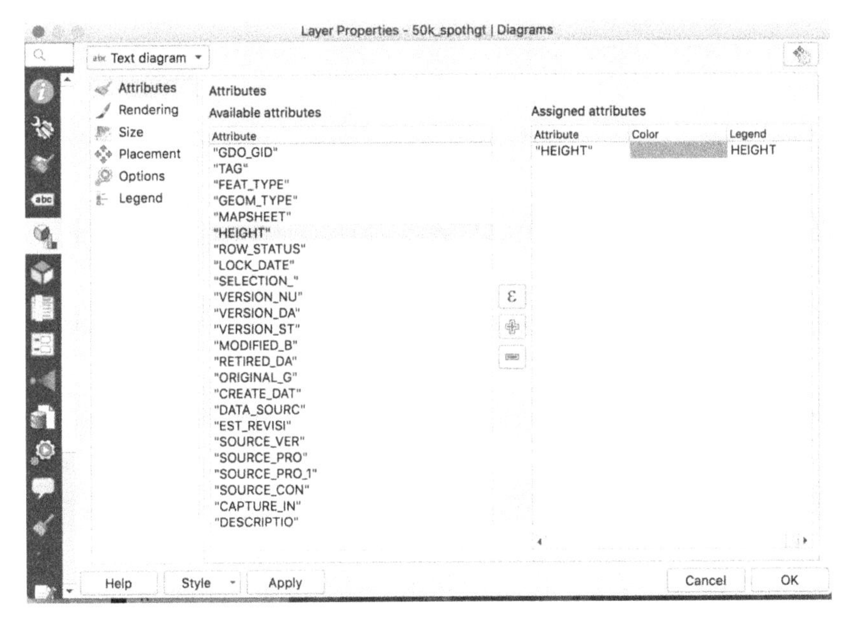

Click *Apply* and you will see diagrams added to the map showing the height of each point:

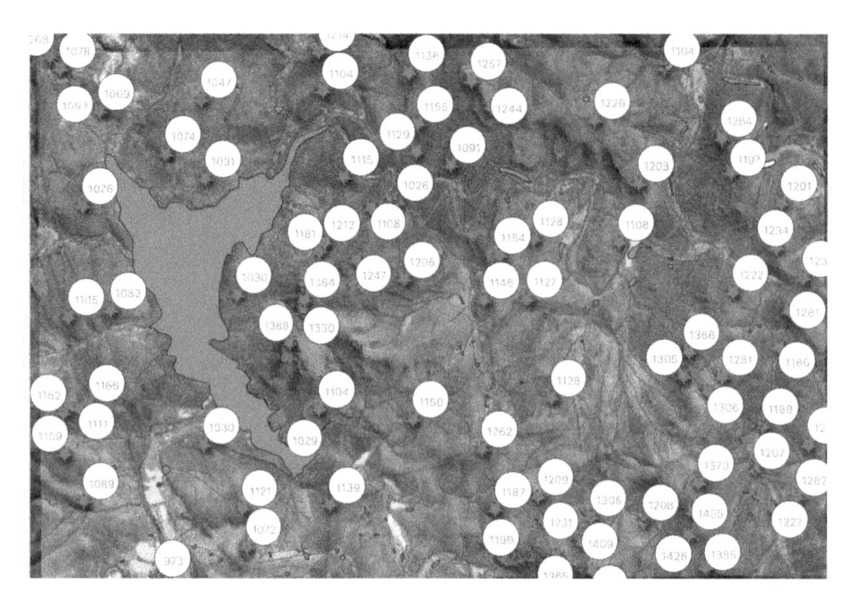

You could add another or several attributes for a pie chart for area data. This works well with pie charts of data representing percentages and amounts.

To alter the size, color, etc. use these options, as shown below:

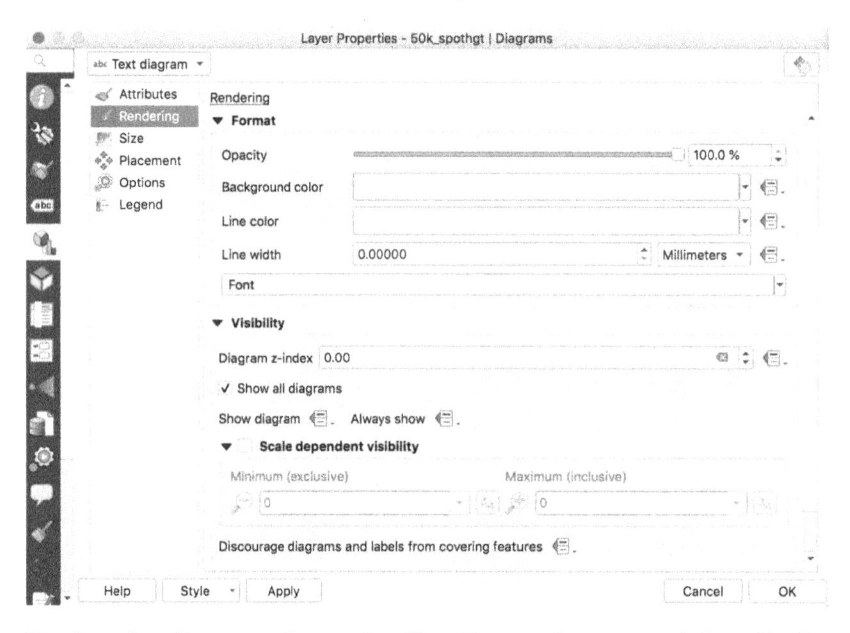

To clear the diagram, change the *Text Diagram* box at top left to *No Diagram*, or click the green minus icon to remove the attributes that are chosen.

2.28 Statistical Summary

You can view a quick statistical summary of data by clicking on the Sigma icon $\sum$. A statistics pop-up window will open. Enter the information shown below, for `50k_contours` and the *HEIGHT* attribute, to see the data for the height attribute, as shown below. You can see that there are 212 individual contours, ranging from 980 to 1500 meters above sea level, along with the mean, minimum, max, etc. of the contours. Some of these numbers are meaningless, like the sum, depending on the data.

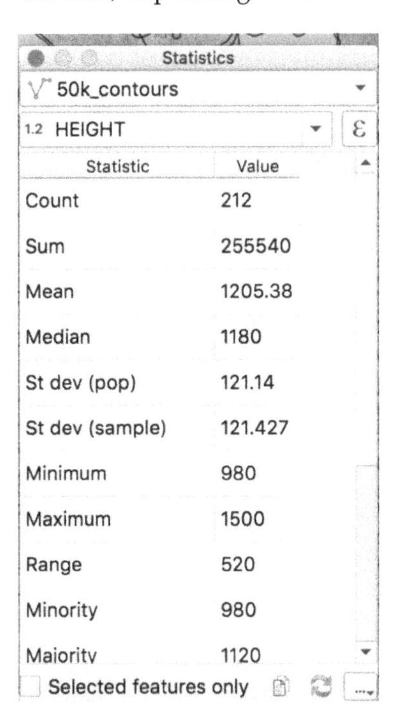

Note that at the bottom you can work with selected features only, and you can send the contents to the clipboard to copy and paste the data elsewhere.

2.29 Query of vector features

We will now work some with the vector topographic contour file `50k_contours`. They may be difficult to see, so make the contours dark blue or black so we can see them better. Simply double-click on the horizontal line next to the name in the *Layers* panel and you will be taken to the *Symbology* dialog. Click on the color bar and pick dark blue or black and then click *OK*. Notice the large number of vector line styles available.

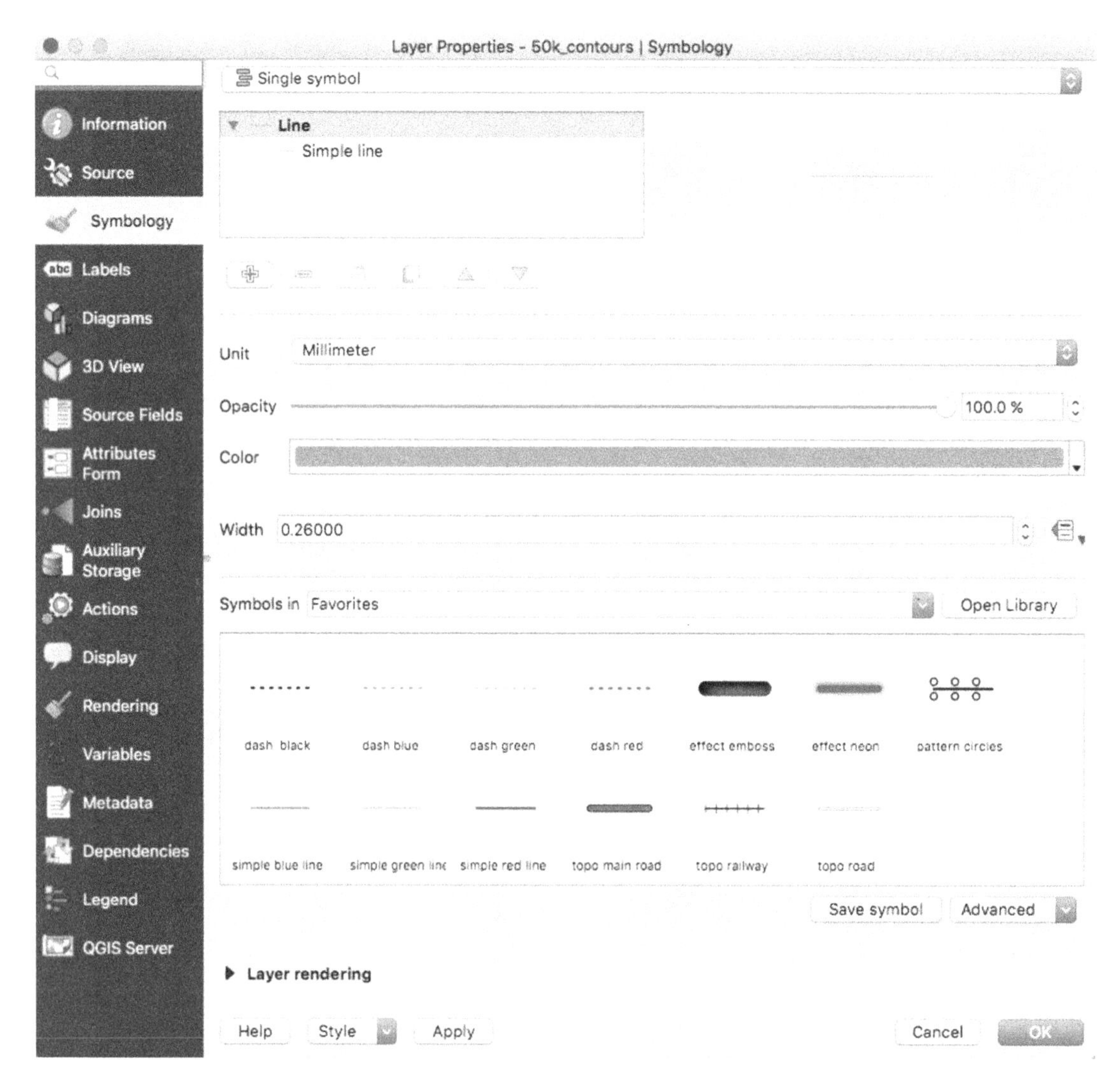

Now that you can see the contour lines better, click on the layer name in the *Layers* panel it is highlighted, and then click the *Identify Features* icon and click on several contours. It will highlight the contour and show you the contents of the database.

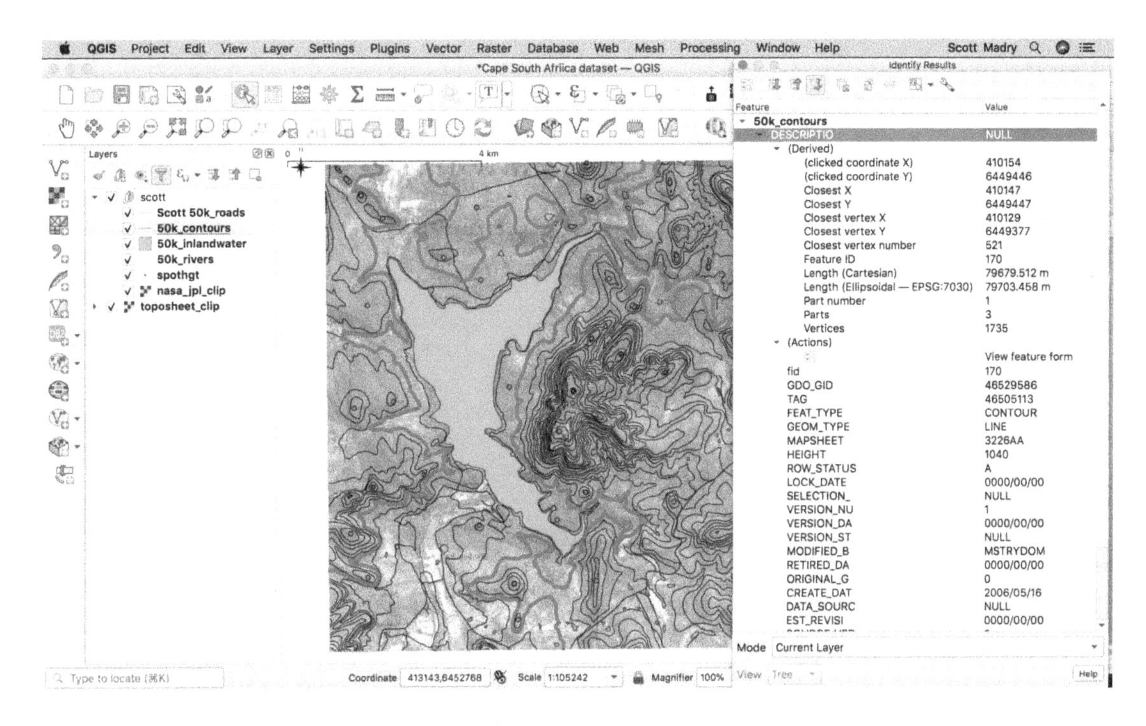

2.30 Vector Query Builder

Click on several contours to get a range of their elevations (around 1040-1360 meters). You could also look the attribute file or use the statistical summary icon.

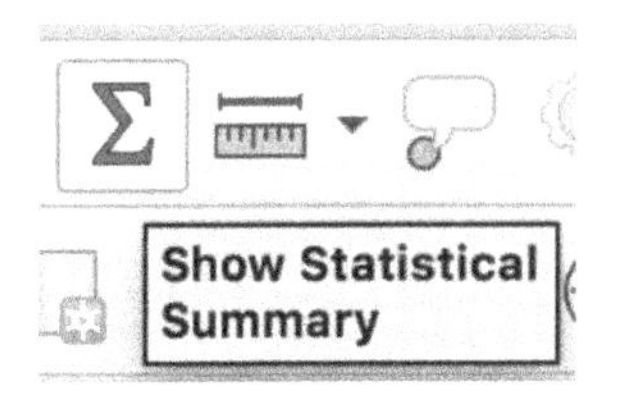

Right-click on the contour layer in the *Layers* panel, and click on `Filter`. This opens *Query Builder* dialog. It is very easy to use. Double-click on *HEIGHT* under fields at left, then >=, and type 1300. Click on *OK* at bottom right to filter the layer:

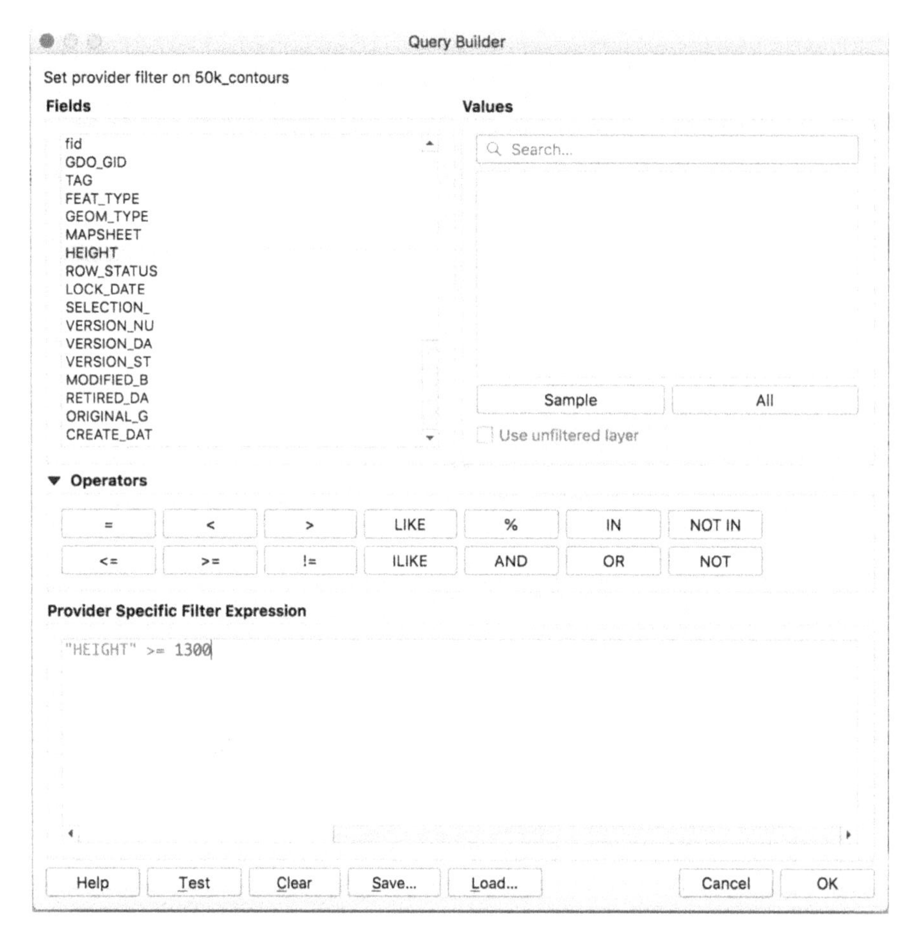

You should see this, with only the contour lines above 1300 meters displayed.

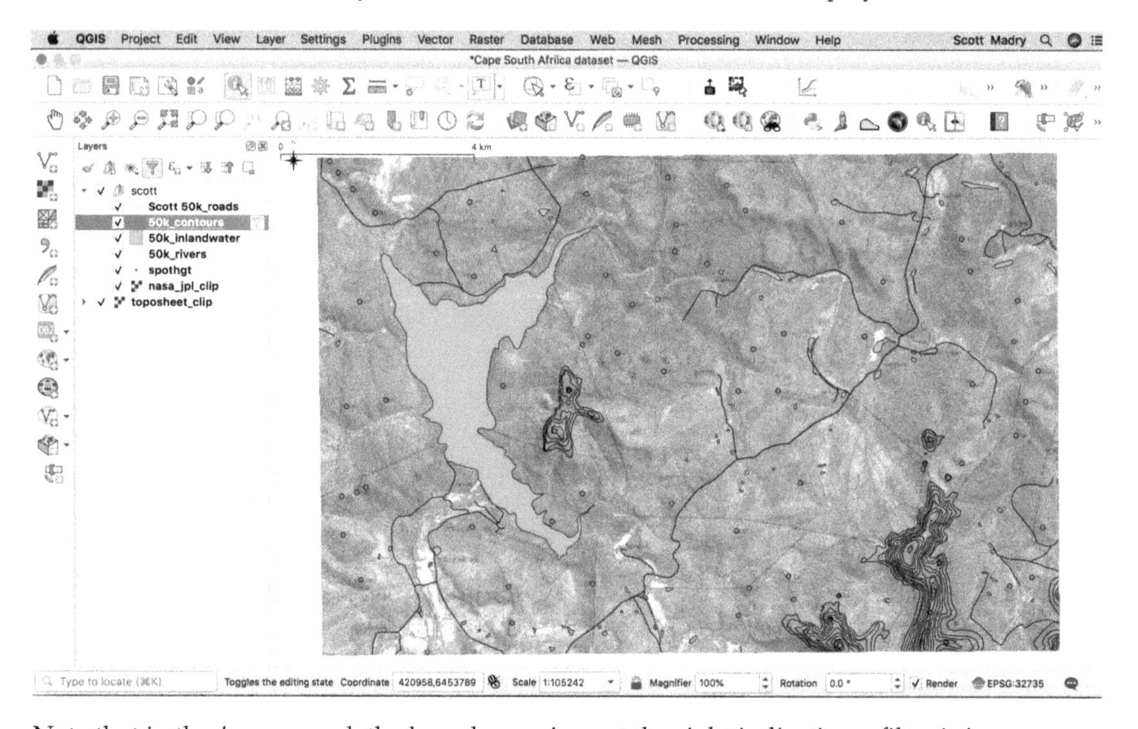

Note that in the *Layers* panel, the layer has an icon at the right indicating a filter is in use:

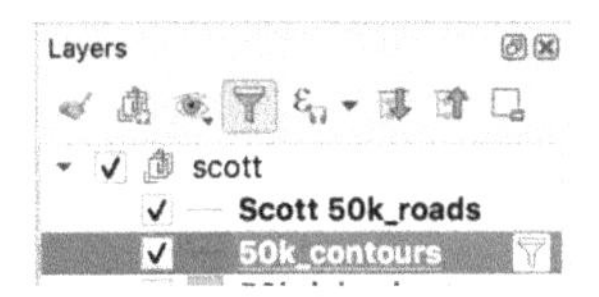

If you click on the filter icon, it opens the filter menu. Try some other combinations, such as shown below:

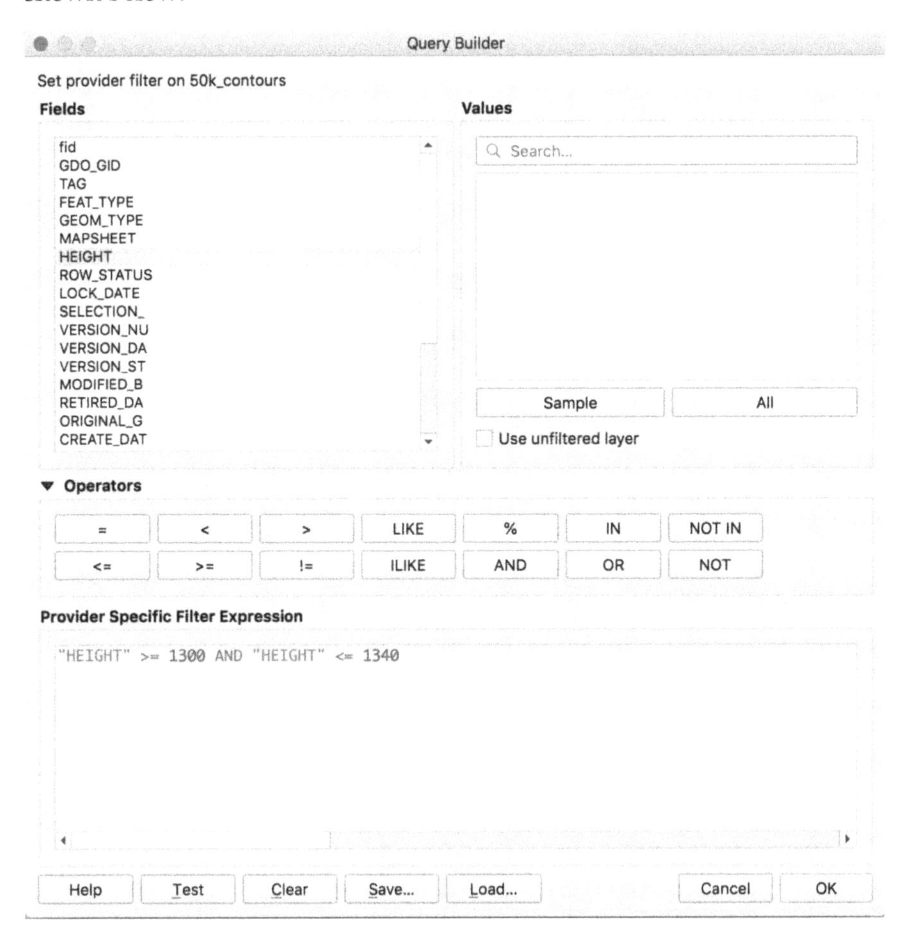

This filter expression will display only contours between 1300 and 1340 meters. You can save your selected features as a new shape file by right-clicking on the file in the *Layers* panel and selecting `Export->Save Features As`. Be sure to go back and clear the filter when you are finished with it, by clicking on the *Clear* button and clicking *OK* on the *Query Builder* dialog. You will now see all the contours. If the filter is left in place, any GIS operations with that layer will ONLY use or display the selected data. You can try it with other vector point, line, or polygon data as well.

2.31 Complex Vector Representations

You can manipulate the vector representation by right-clicking on the file in the *Layers* panel and clicking on `Properties`. Do this for the `50k_contours` layer. Under *Style*, click on the *Single Symbol* tab at top left, and change it to *Graduated*.

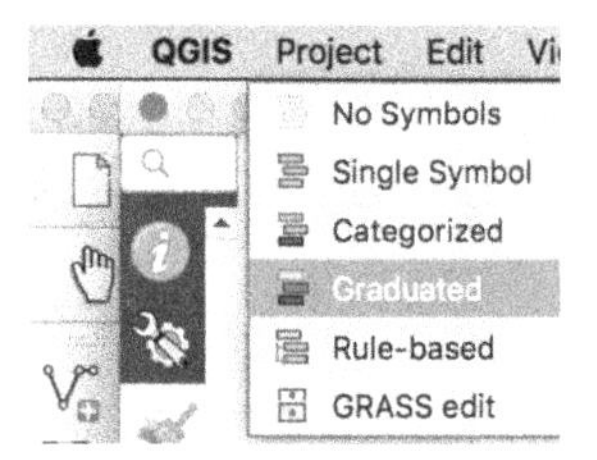

Then change the value to *HEIGHT* and click *Classify* as shown below:

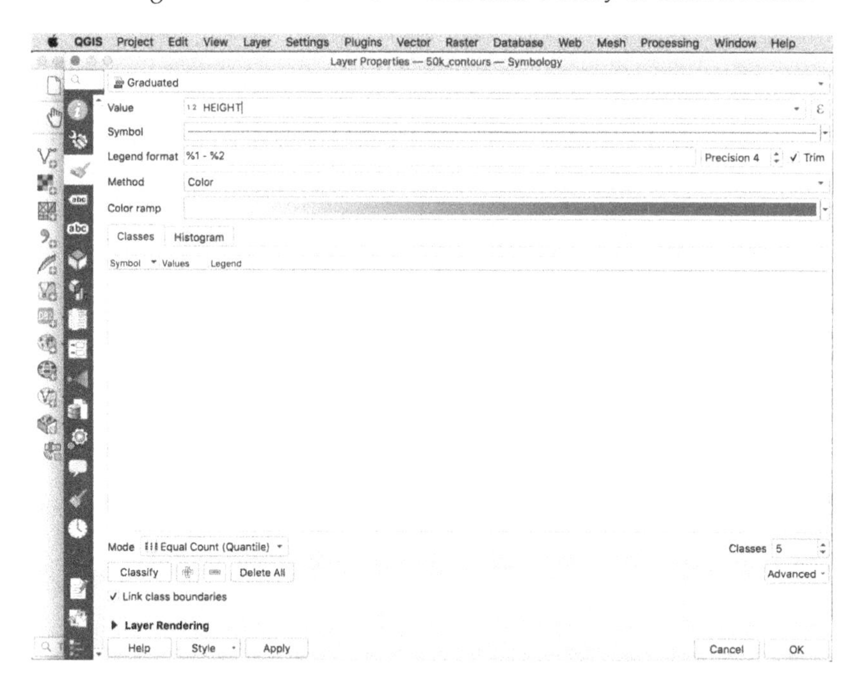

This will separate the vector contours into five equal interval classes with different colors. You can change the number of classes by clicking the up/down arrow at top right Classes 5. Now click on each symbol (the line between the check mark and the value) and assign them a different color and line width. Click *OK* and you should see your contours grouped into five equal classes, as shown below. Note that your classes with the contour ranges are now displayed in your *Layers* panel. You can change the colors as you wish.

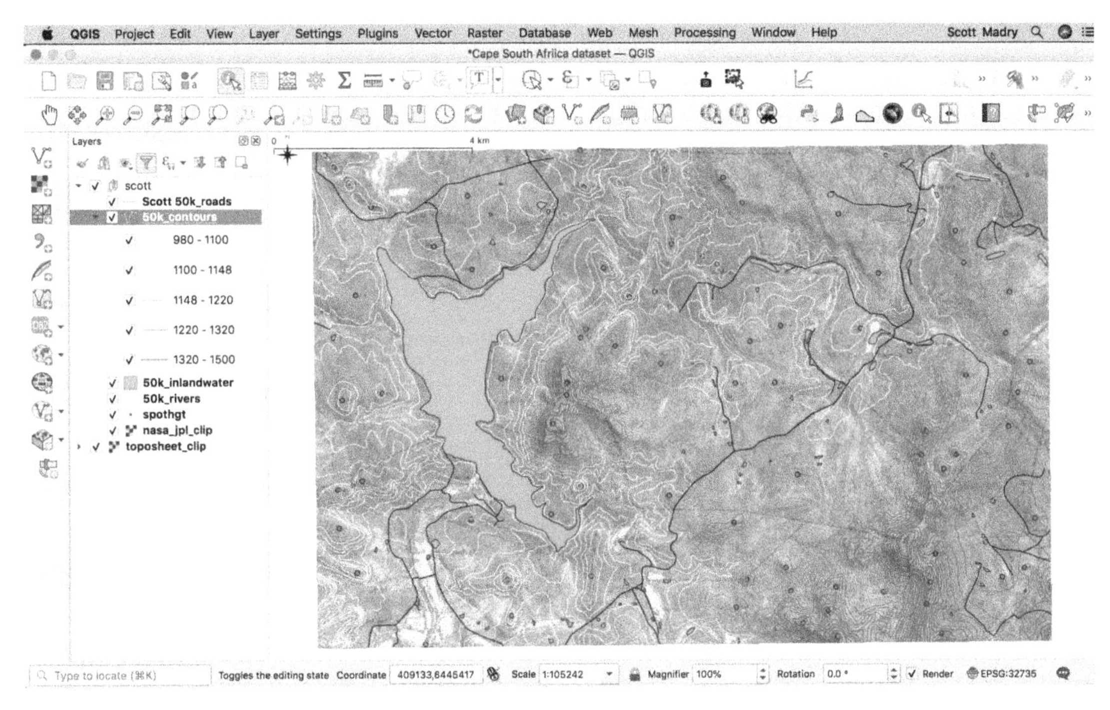

Note that equal intervals are not your only option:

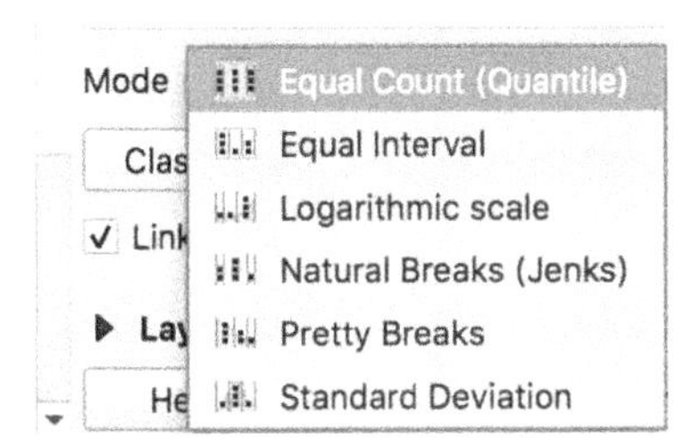

Try doing the same with the point layer `spothgt`. First look at the attribute table by right-clicking on the layer in the *Layers* panel at left and choosing `Open Attribute Table`:

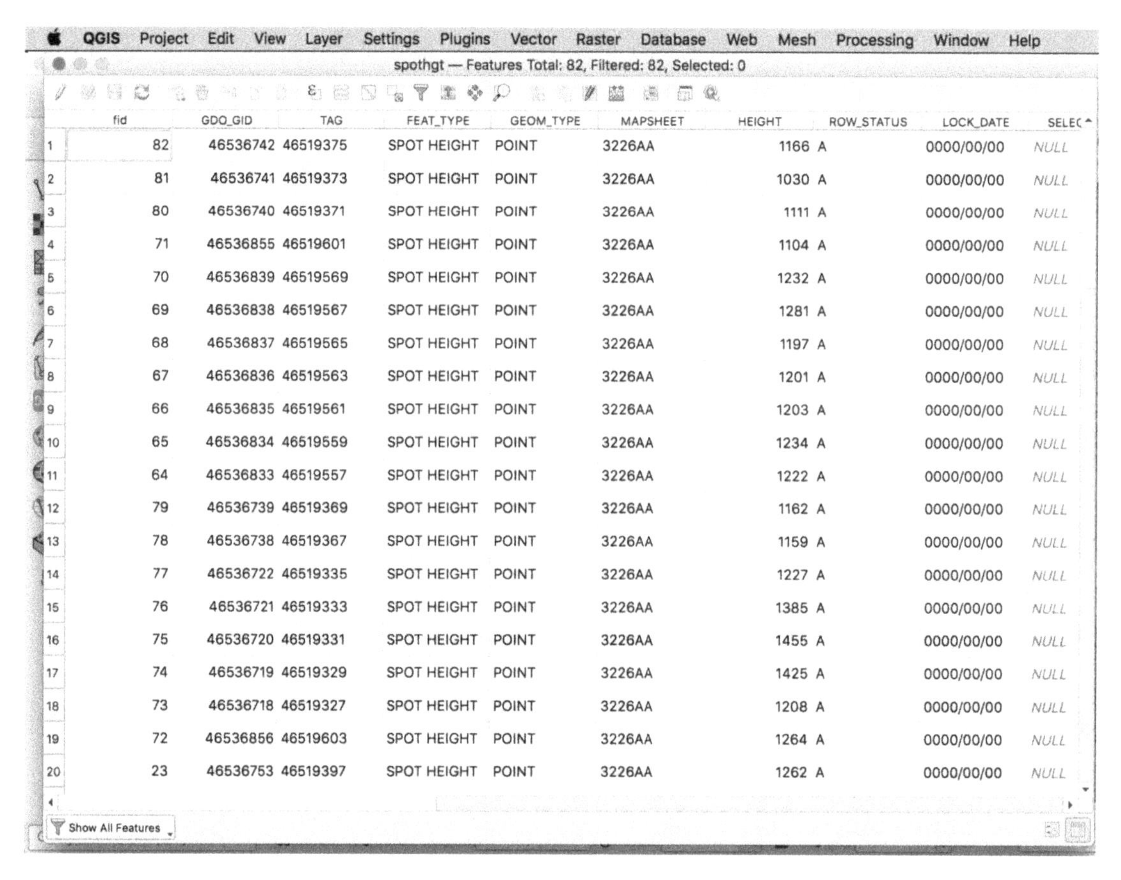

	fid	GDO_GID	TAG	FEAT_TYPE	GEOM_TYPE	MAPSHEET	HEIGHT	ROW_STATUS	LOCK_DATE	SELEC
1	82	46536742	46519375	SPOT HEIGHT	POINT	3226AA	1166	A	0000/00/00	NULL
2	81	46536741	46519373	SPOT HEIGHT	POINT	3226AA	1030	A	0000/00/00	NULL
3	80	46536740	46519371	SPOT HEIGHT	POINT	3226AA	1111	A	0000/00/00	NULL
4	71	46536855	46519601	SPOT HEIGHT	POINT	3226AA	1104	A	0000/00/00	NULL
5	70	46536839	46519569	SPOT HEIGHT	POINT	3226AA	1232	A	0000/00/00	NULL
6	69	46536838	46519567	SPOT HEIGHT	POINT	3226AA	1281	A	0000/00/00	NULL
7	68	46536837	46519565	SPOT HEIGHT	POINT	3226AA	1197	A	0000/00/00	NULL
8	67	46536836	46519563	SPOT HEIGHT	POINT	3226AA	1201	A	0000/00/00	NULL
9	66	46536835	46519561	SPOT HEIGHT	POINT	3226AA	1203	A	0000/00/00	NULL
10	65	46536834	46519559	SPOT HEIGHT	POINT	3226AA	1234	A	0000/00/00	NULL
11	64	46536833	46519557	SPOT HEIGHT	POINT	3226AA	1222	A	0000/00/00	NULL
12	79	46536739	46519369	SPOT HEIGHT	POINT	3226AA	1162	A	0000/00/00	NULL
13	78	46536738	46519367	SPOT HEIGHT	POINT	3226AA	1159	A	0000/00/00	NULL
14	77	46536722	46519335	SPOT HEIGHT	POINT	3226AA	1227	A	0000/00/00	NULL
15	76	46536721	46519333	SPOT HEIGHT	POINT	3226AA	1385	A	0000/00/00	NULL
16	75	46536720	46519331	SPOT HEIGHT	POINT	3226AA	1455	A	0000/00/00	NULL
17	74	46536719	46519329	SPOT HEIGHT	POINT	3226AA	1425	A	0000/00/00	NULL
18	73	46536718	46519327	SPOT HEIGHT	POINT	3226AA	1208	A	0000/00/00	NULL
19	72	46536856	46519603	SPOT HEIGHT	POINT	3226AA	1264	A	0000/00/00	NULL
20	23	46536753	46519397	SPOT HEIGHT	POINT	3226AA	1262	A	0000/00/00	NULL

You will see that the attribute *HEIGHT* contains the actual elevation data. Now close the attribute table, and again right-click in the filename in the *Layers* panel, go to `Properties`, again go to *Style*, and at the top, change Single Symbol to Graduated, pick *HEIGHT* for the *Value*, and change the color ramp and classes if you wish, then hit *Classify*.

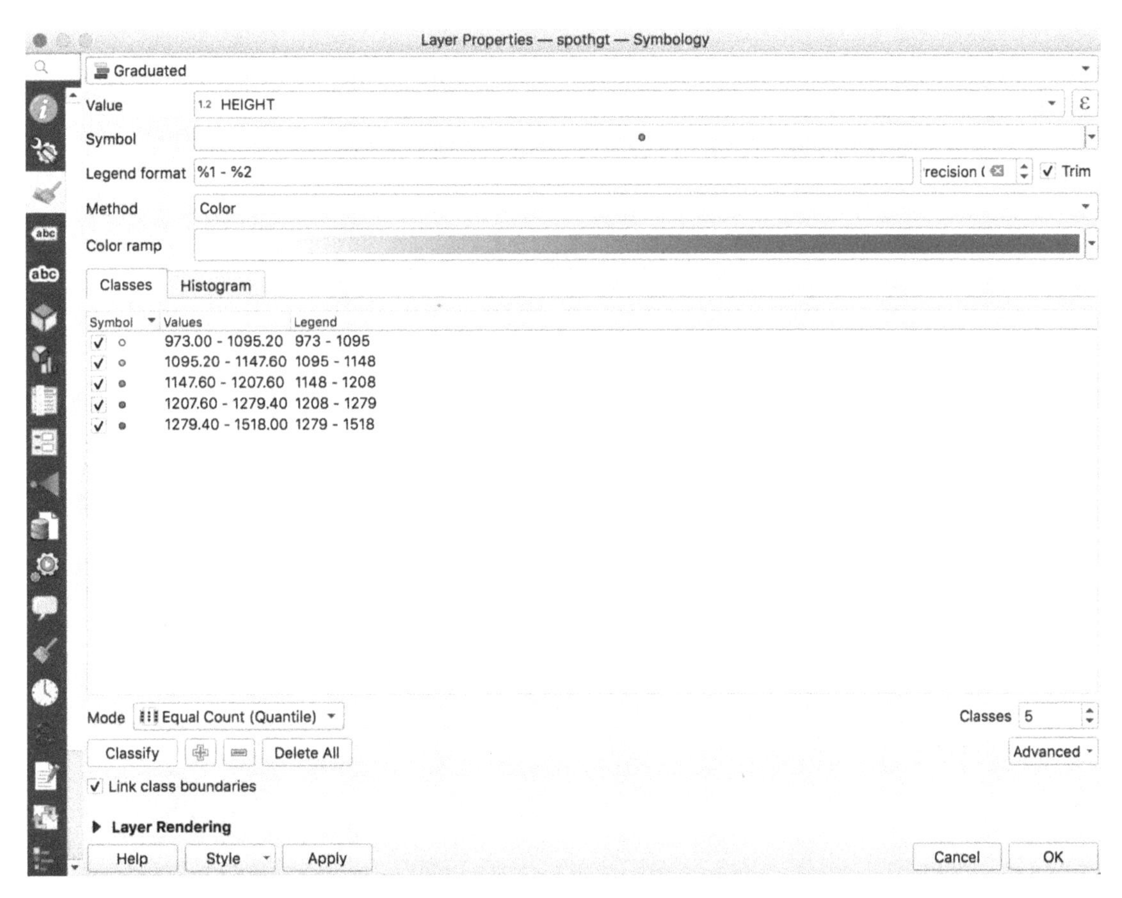

You can now change the display by clicking on each dot under *Symbol*, and you will see that point data also have a different library of symbols (you can add your own as well). Pick a different symbol, color, and size for each class, then hit *OK* and look at your data.

2.32 Vector Data Symbols

QGIS provides multiple ways to symbolize your data:

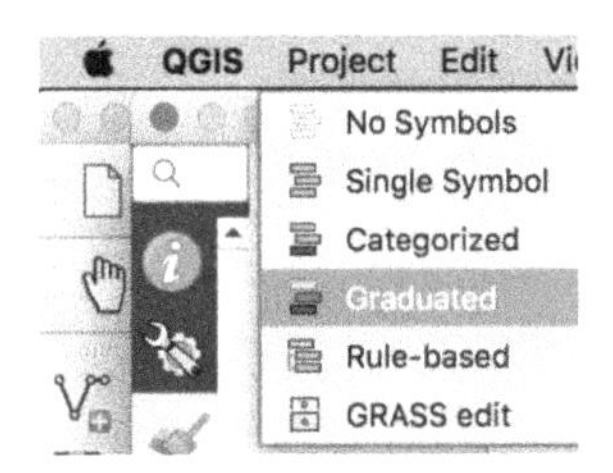

Single Symbol: This uses a single symbol for point layers, a simple outline for line layers, and a fill color for polygons. All features in the layer are rendered the same way.

Categorized Symbol: This allows you to symbolize features based on an attribute value. For example, you could categorize by bird name, age, etc. Each category is rendered with the same color/symbol.

Graduated Symbol: This renderer allows you to symbolize features based on a number of classes. Each class is composed of a range of values. The field used to render the features must be numeric.

Rule-Based: This allows you to create specific rules for the styling groupings. This is a very powerful capability, and you can define the rules using simple SQL language provided in QGIS.

GRASS Edit: This allows you to edit the symbology for GRASS format vector files.

The choice of renderer depends on what you are trying to visualize. To learn more, click on *Help* at the bottom left of the vector layer properties dialog box.

2.33 Vector Data Properties

In the `Properties` menu, accessed by right-clicking on a layer name in the *Layers* panel and choosing `Properties`, you have many of the common GIS background functions. Note that the `Properties` options will be different for vector (point, line, polygon), and raster data. For vector layers, under *Information* at top left, you have the data layer name, source, coordinate reference system.

From the *Rendering* tab you can set a *Scale Dependent Visibility*. This allows you to have data of a small scale (things look small) only be displayed if you are zoomed out to a large area. Likewise, you can have detailed data (large scale) only "turn on" and be visible when you are zoomed in. You can also simplify the vector geometry or force layers to render as a raster.

Under *Symbology* you can alter the color, fill (for polygons) and width of the lines. Under *Labels* you can enable text labels and set the size, resolution of zoom, etc. Note that the labels for the layer `50k_spothgt` has the *HEIGHT* turned on, so you see the elevation number by each point icon.

FIELDS shows you the attributes, The *Actions* tab allows you to do external actions such as opening a word processing document, or do a web search on a layer name. On the *Joins* tab you can specify a vector join of new attribute data. *Diagrams* lets you display diagrams, pie charts, etc as we illustrated previously.

2.34 Metadata

Metadata is very important. You should be sure to fill in and update metadata (information about your data), especially when you create or edit layers. People often ignore this, but it is very important. Please always document your data, for your own use and for others who may use your data. Here is a blank *Metadata* tab to be filled in:

Note the row of tabs along the top:

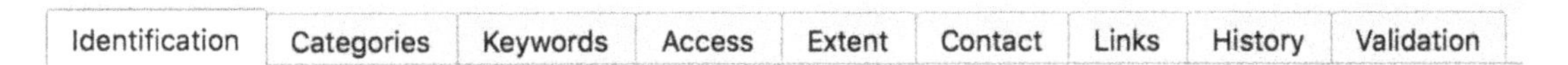

Each of these have relevant categories to be filled out. Metadata can be automatically entered if standard metadata, such as U.S. FGDC-CSDGM standard metadata. To learn more about metadata, standards, and best practices see: `https://www.usgs.gov/products/data-and-tools/data-management/metadata-creation`

2.35 Raster data properties

Raster data properties include *Information* and *Symbology*, as well as *Transparency*, *Pyramids*, *Histograms*, and *Metadata*. *Information* is the same as for vector data, but *Symbology* lets you render raster images using various color, gray scale, and palette color tables. The *Transparency* tab lets you make an image partly transparent, and *Pyramids* is a raster data compression capability.

2.36 Raster Data Pyramids

Be aware of these warnings:

> Please note that building internal pyramids may alter the original data file and once created they cannot be removed!
> Please note that building internal pyramids could corrupt your image - always make a backup of your data first! Raster Histograms.

2.37 Histogram

The *Histogram* tab lets you view the RGB distribution of values in your image. Here is a histogram of the satellite image `nasa_jpl_clip`, showing the frequency of distribution of pixel values for red, blue, and green:

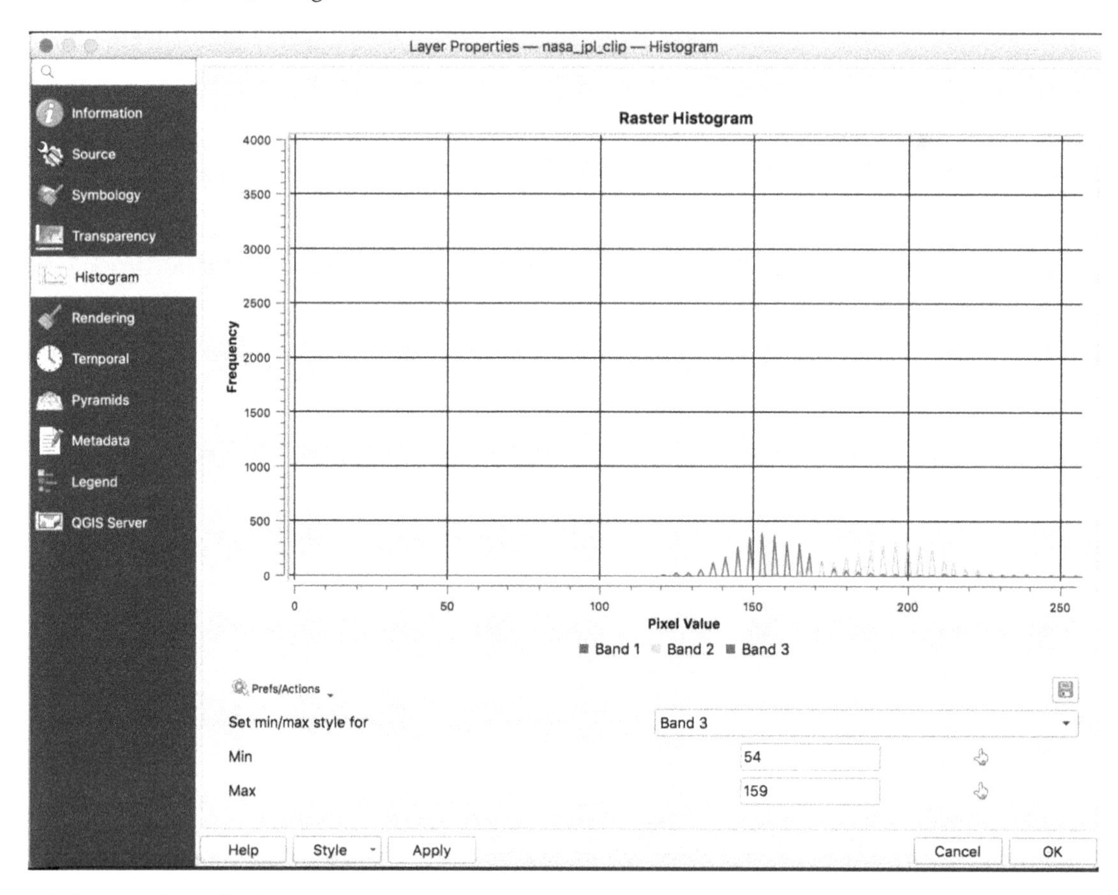

Click on *Help* at the bottom left to get more information on any QGIS module. Here is the help page for histogram:

15.1.5. Histogram Properties

The **Histogram** tab allows you to view the distribution of the values in your raster. The histogram is generated when you press the **Compute Histogram** button. All existing bands will be displayed together. You can save the histogram as an image with the button.

At the bottom of the histogram, you can select a raster band in the drop-down menu and **Set min/max style for** it. The **Prefs/Actions** drop-down menu gives you advanced options to customize the histogram:

- With the **Visibility** option, you can display histograms for individual bands. You will need to select the option **Show selected band**.
- The **Min/max options** allow you to 'Always show min/max markers', to 'Zoom to min/max' and to 'Update style to min/max'.
- The **Actions** option allows you to 'Reset' or 'Recompute histogram' after you have changed the min or max values of the band(s).

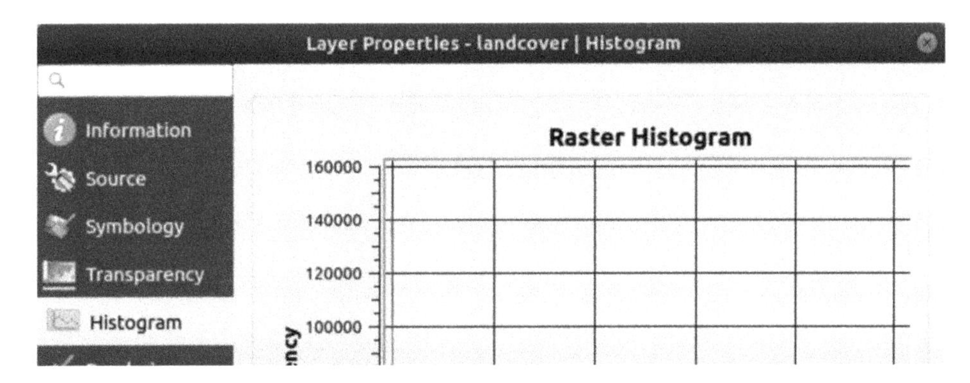

2.38 Loading XYZ tiles into QGIS

The *XYZ Tiles* feature found in the *Browser* panel lets you add various raster web data, including satellite images and maps. To add an XYZ tile layer into your current QGIS workspace, right-click on *XYZ Tiles* in the *Browser* panel and select `New Connection`.

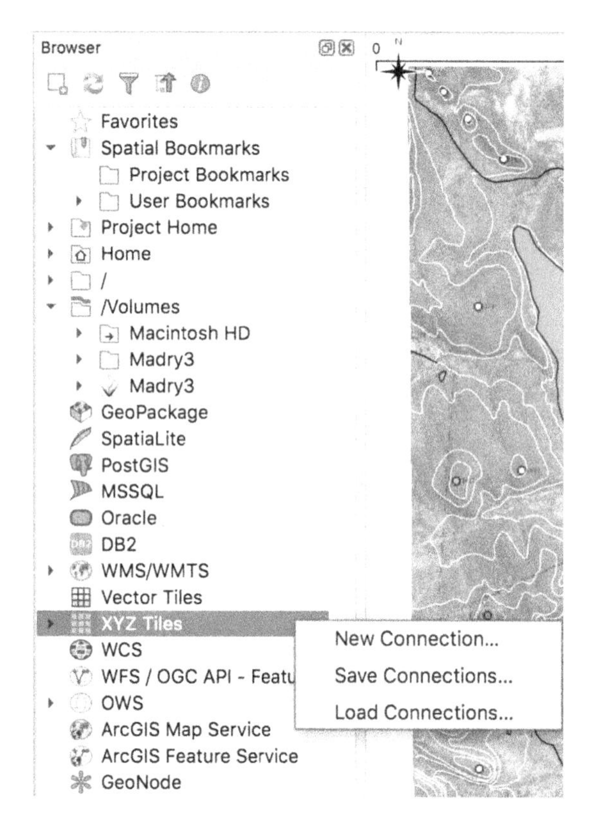

You will see a dialog box, where you can enter the URL of a data server:

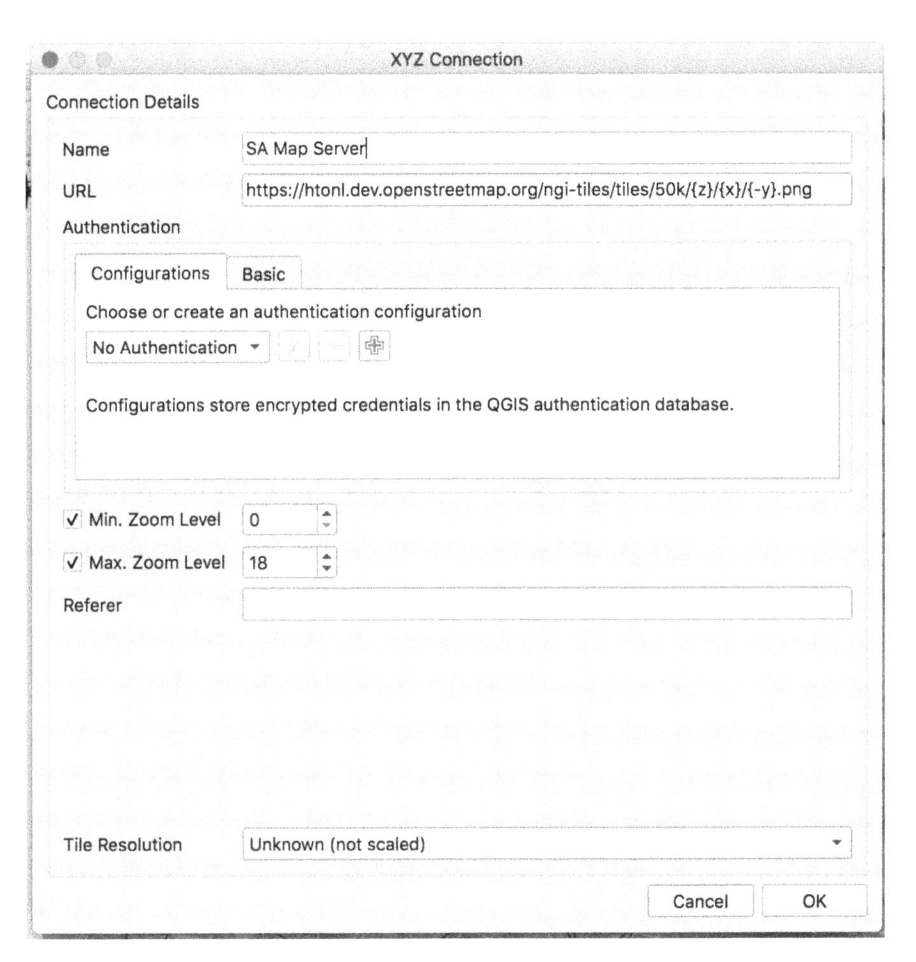

Cut and paste the following URL for the 1:50,000 topographic maps of South Africa into the URL above, and give it a name, like "South Africa maps", and click *OK*: `https://htonl.dev.openstreetmap.org/ngi-tiles/tiles/50k/{z}/{x}/{-y}.png`

You will now see "South Africa maps" in your *Layers* panel. You can view all your XYZ tiles by clicking on the small triangle at left to expand the list.

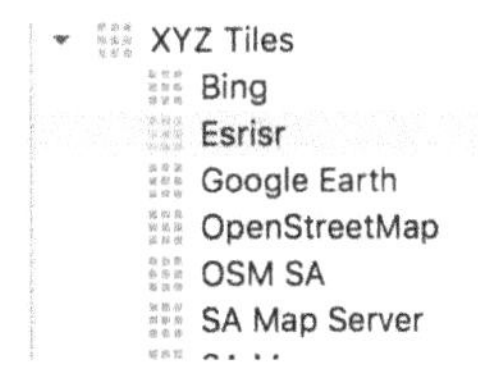

Double-click on the "South Africa maps" in the tile list and it will be added to the map. Here is a URL for adding an imagery file for South Africa NGI Imagery: `http://aerial.openstreetmap.org.za/ngi-aerial/{z}/{x}/{y}.jpg`

One benefit is that these data are not stored permanently on your computer, and you can zoom and pan and the maps will "follow" you. There are many such servers available. The QuickMapServices plugin (discussed below) offers another simple way to access many of the major such sources, but you can always use the *XYZ Tiles* feature in your *Browser* panel as well. Visit this site to see many more: `https://github.com/nextgis/quickmapservices_contrib/tree/master/data_sources`

You can turn these data on and off or make them transparent, and you can remove the layer from your project by right-clicking on the name in the *Layers* panel and then clicking `Remove Layer`, as shown below:

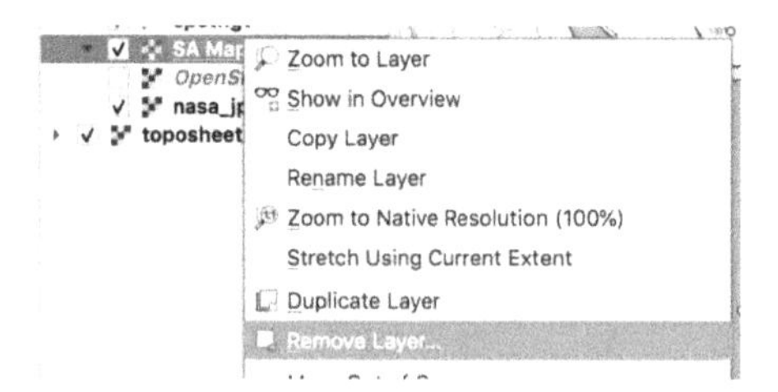

2.39 Preview Mode

In the `View` menu you will find `Preview Mode`. This allows you to view your data in gray scale (very useful for publications that do not allow color images). Also available are options for color blind color options.

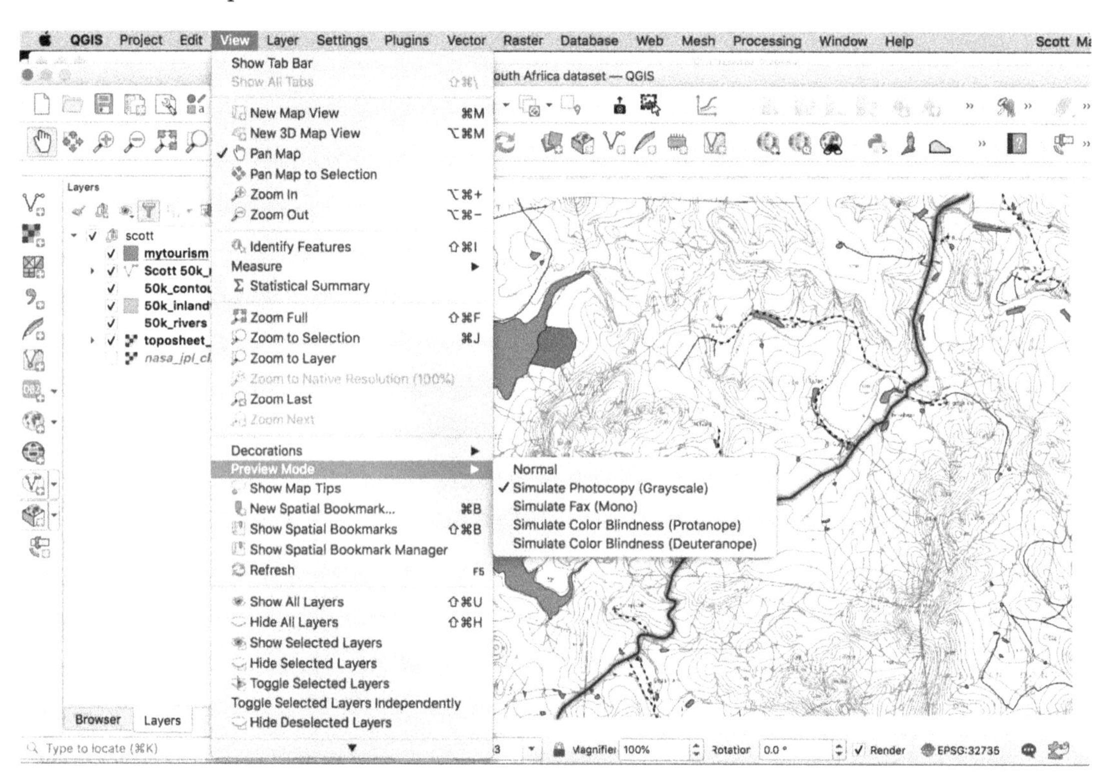

There are many other capabilities available, so take a look at all the QGIS menu options.

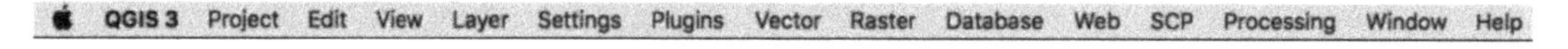

`Help` at the far right provides access to a search option where you can enter a module name, etc, as well as access to the QGIS help contents, the QGIS home page, and how to report an issue or bug and even find commercial support.

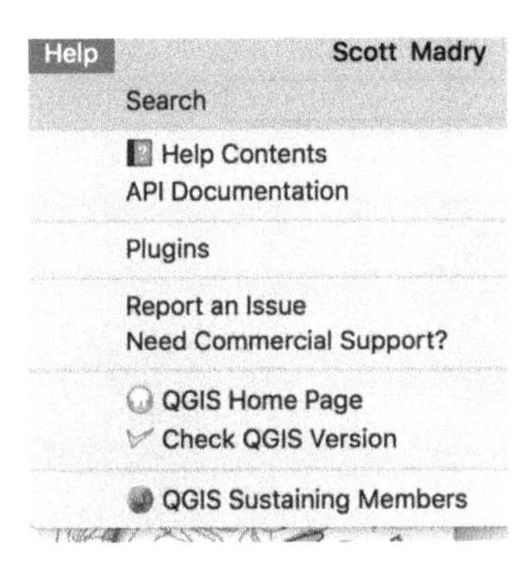

This is only the very beginning of the capabilities of QGIS, and so next we shall look at creating and editing new vector GIS data.

3. Creating and Editing New Vector Files in QGIS

3.1 Creating New Vector Data

While it is simple to load an existing vector shapefile, GeoPackage file, or other file into QGIS, we often want to create our own data. It is simple to create a new vector shapefile or GeoPackage file in QGIS representing points, lines, or polygons. One of the major differences between QGIS and ArcGIS is that pretty much everything in QGIS, including vector editing, is done right in the standard user interface. We will create a new polygon layer that outlines a boundary for the tourism hiking area to the east of the lake. Open the vector `tourism_polys.shp` file, and then click it off and on to see where it is, and zoom into that area. From the top toolbar select `Layer->Create Layer->New Shapefile Layer`.

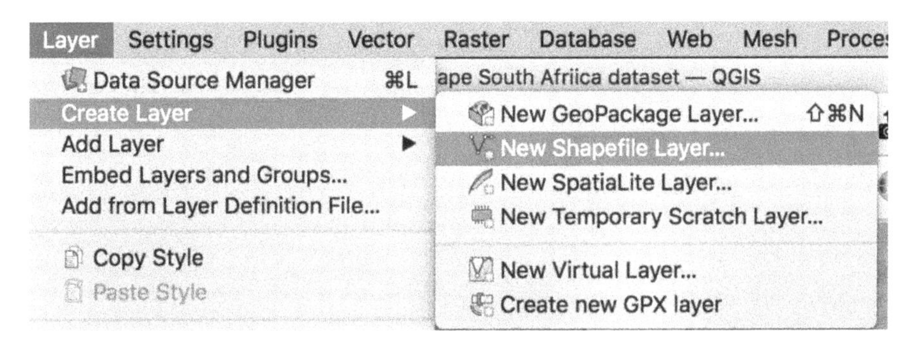

Note that you can also create other file formats, as seen in the `Create Layer` menu. It is recommended to use GeoPackage format, but here we will create an ESRI Shapefile.

You will see a new dialog box—select the type of file you want to create (single or multi point, line, or polygon) under *Geometry* type. We want to define an area, so we need polygon geometry. Specify the CRS (WGS84 UTM35S). Notice that the project's CRS is available for quick selection:

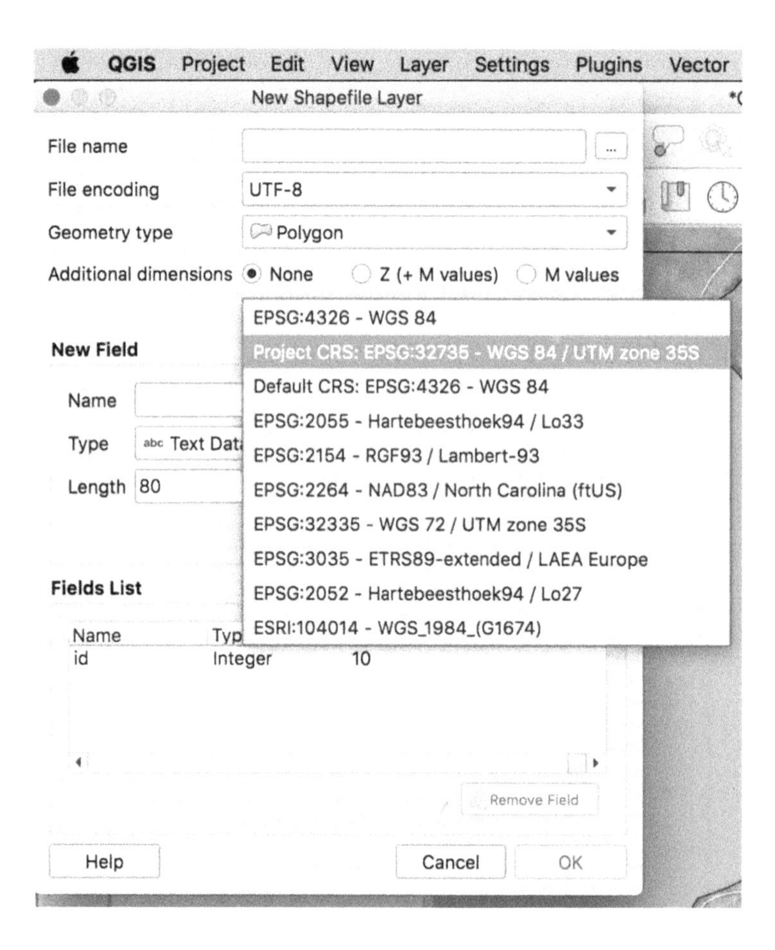

3.2 Creating New Vector Data Attribute Fields

Add the attributes you want by entering at least the *Name, Type* in the *New Field* section. You can specify *Length* and *Precision* as appropriate for the field *Type*. Click *Add to Fields List* to add the attribute field. In this case we will make a new text field, and call it *field1*.

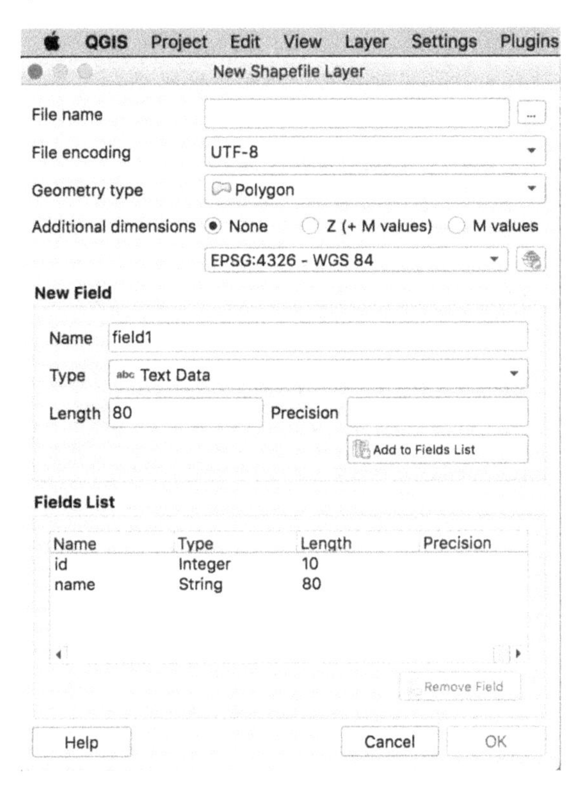

Choose a name and location for the shapefile—same as your other GIS data would be best. Add several new fields for the attribute table, choosing from whole numbers, dates, etc. Call it `mytourism` (or whatever you want) and click *OK* to create the shapefile. Remember to document where you put it.

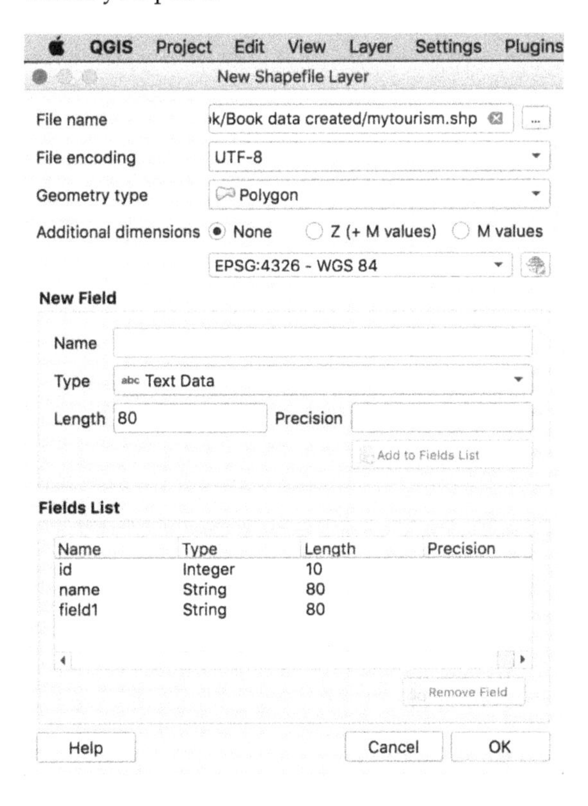

It will automatically be added to the top of the *Layers* panel on the left of your screen.

3.3 The QGIS Editing Mode and Icons

To edit the new shapefile, right-click on it in the *Layers* panel and click on *Toggle Edit* in the popup menu. You can also click on the layer to highlight it, then click the *Toggle Edit* icon in the toolbar. Now additional editing tools in your icon list are no longer grayed out, making them available for use.

You can hover over each icon to see what they do. These are the basic editing functions. There are additional ones under the `Edit` menu.

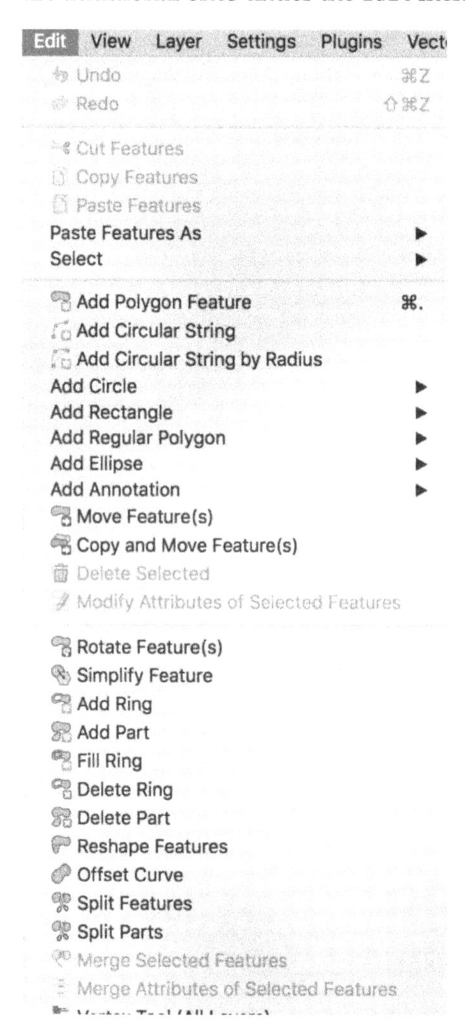

> You may have noticed that not all the tools in the `Edit` menu are present on the toolbar. These are found on the *Advanced Digitizing Toolbar* which you can enable from the `View->Toolbars` menu.

Click the `Toggle Editing` tool and select the `Add Polygon Feature` tool from the editing toolbar to start digitizing with your mouse or track pad. Draw your study area around the hiking area using your mouse. First click on one corner, and then click (or drag) your mouse to another location to create a new, straight line. Continue adding vertices to the polygon and you will see that the it is displayed as you click or drag. When you are ready to close the polygon (your last point), right-click to finish the polygon. You will be prompted to fill in the attribute information:

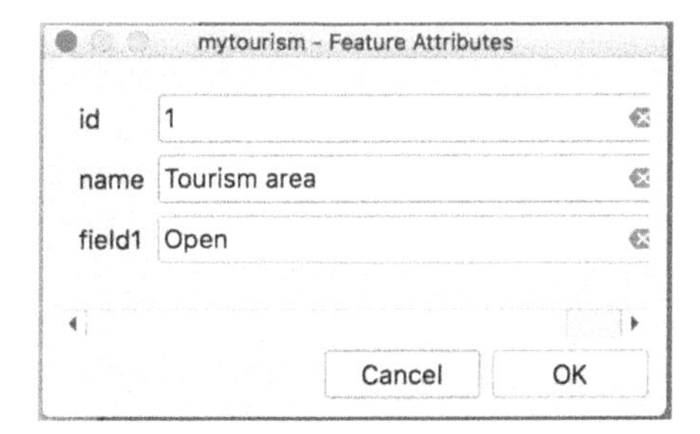

Click *OK* and your new polygon will be displayed. This is what it should look like:

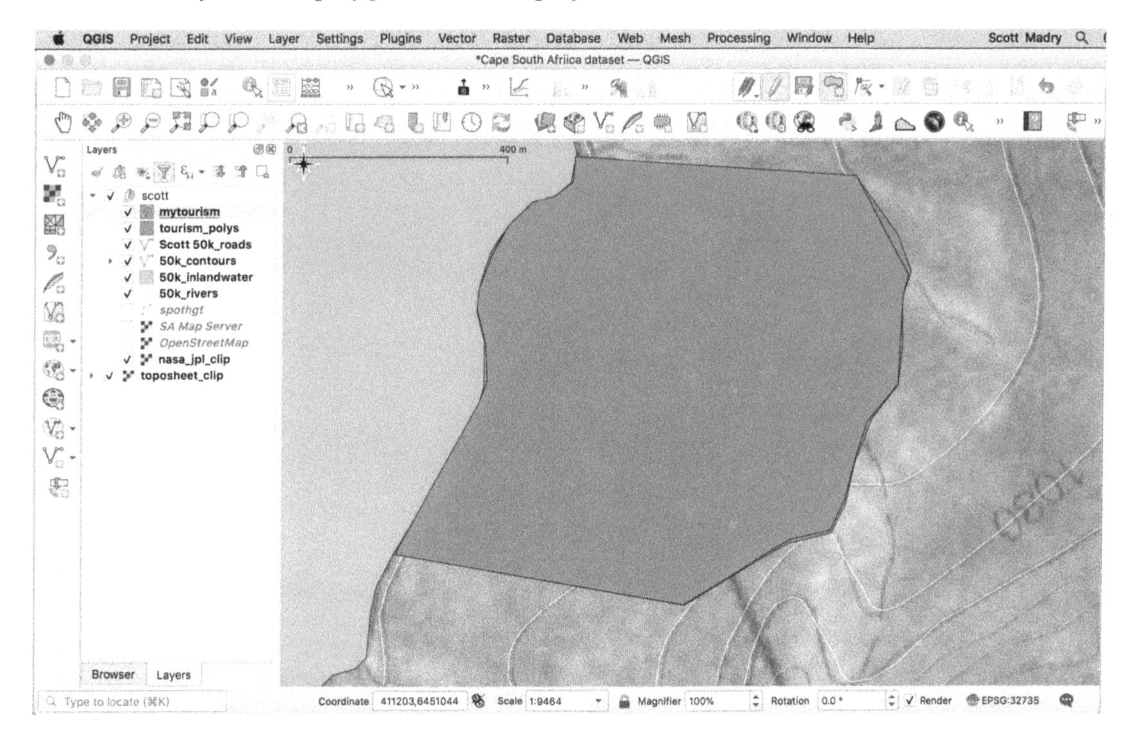

You can right-click on the new layer name at left (`mytourism`) and look at the attribute table, and you can go in and edit the attributes and add new ones. Be sure to save your project and always fill out the metadata for the new layer. You will likely not remember weeks or months from now what you did. The metadata file provides you with a complete way to document your data. Visit each of the tabs across the top of the metadata form to fill out the relevant information:

If you click on the `Edit` menu you will see the complete list of editing options available. Some features may be grayed out depending on where you are in the editing process:

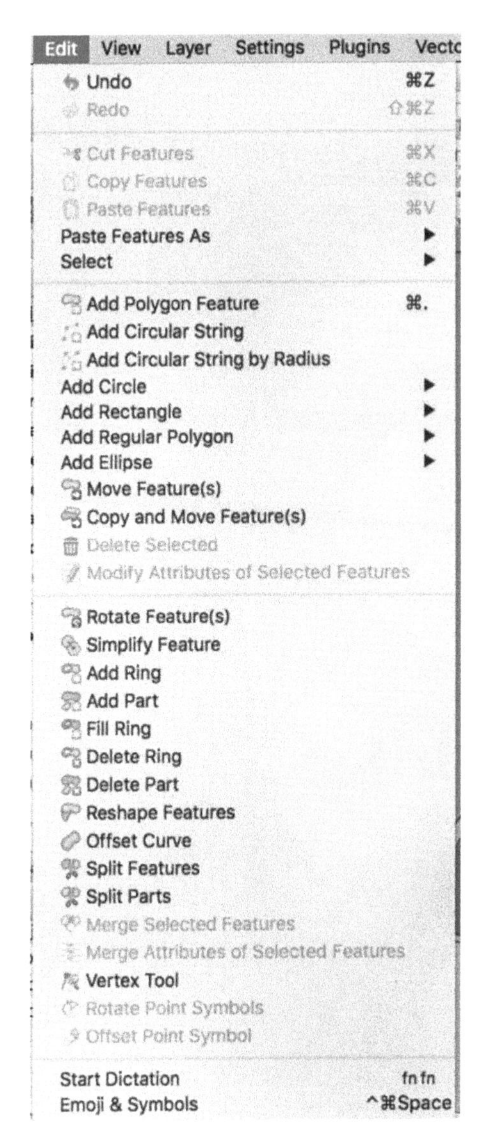

Make the vector file editable again, and experiment with moving the area using the move feature icon and editing the nodes using the node tool .

Always save your edits. You can always back up a step to undo a mistake. Be sure to save the project, create the metadata for new layers, and document what you did in your work log.

By right-clicking on the layer and selecting `Properties->Symbology` you can edit the symbology. Experiment with different rendering options.

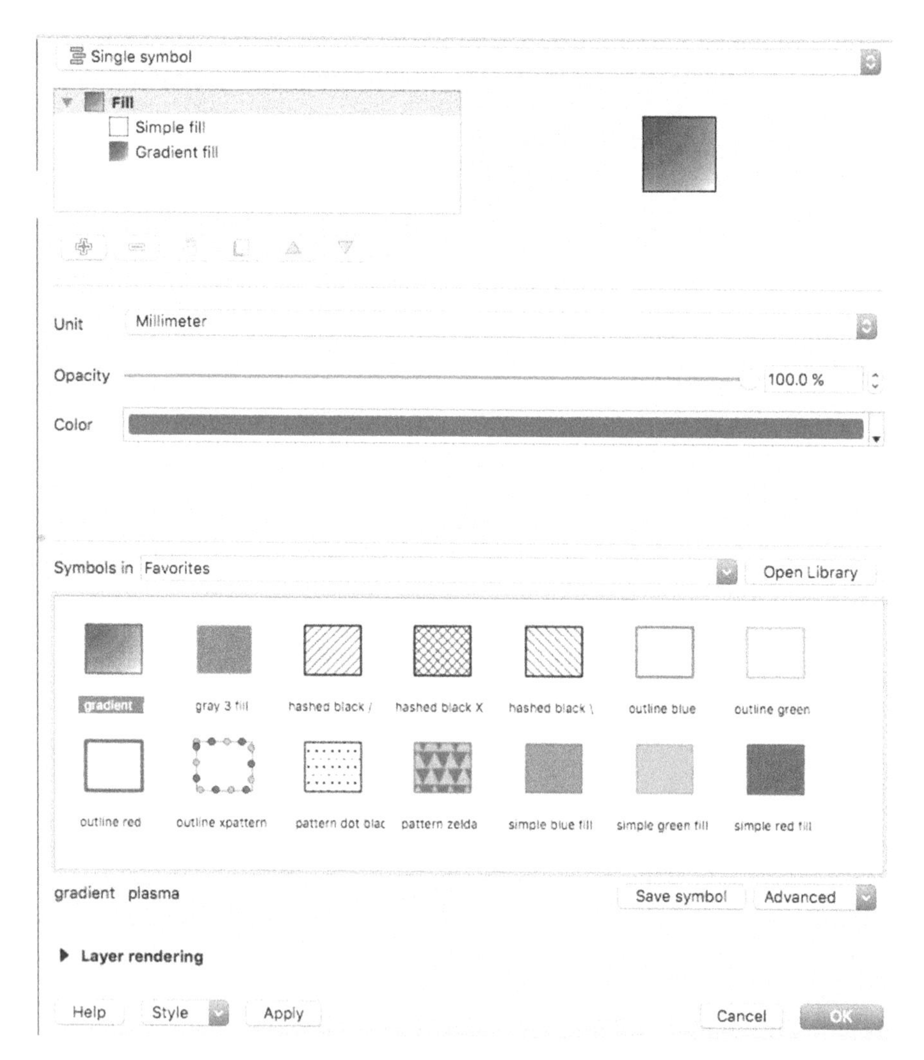

Zoom way in, and see how off you were in digitizing along the coastline. It is important when digitizing that you zoom as far in as you can and be extremely careful. "Garbage in – Garbage out" is an old GIS expression, and your analysis results are only as good as the quality of the data used. Always take great care in creating your data, and always fill out the metadata files as you go so that others (and you later on) know exactly what was done.

3.4 Creating a new GeoPackage Layer

To create a new GeoPackage, choose `Layer->Create Layer->New GeoPackage Layer` from the menu and fill in the dialog:

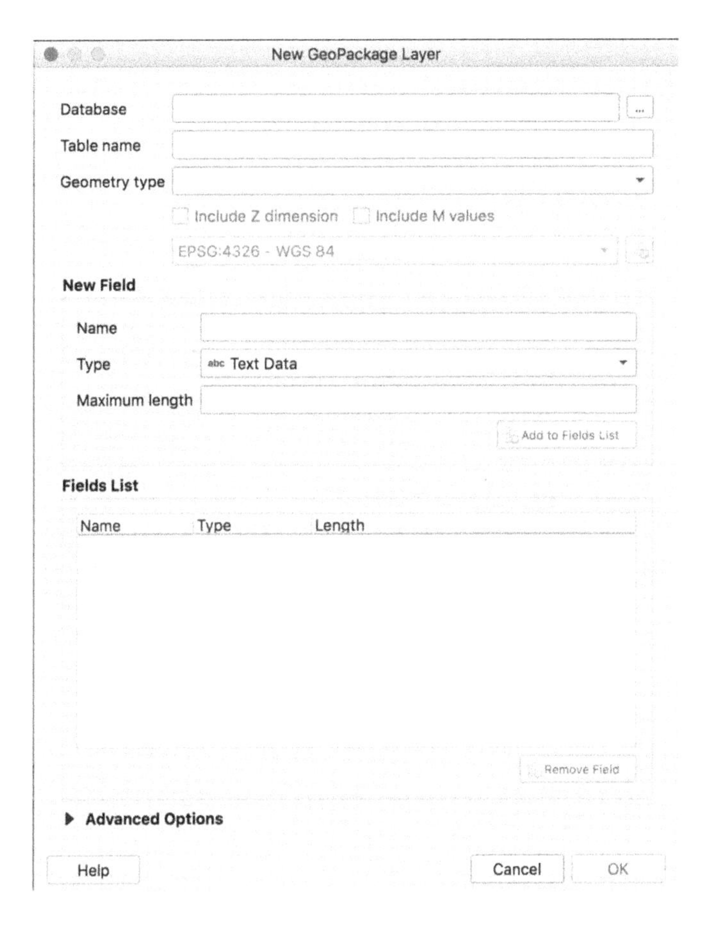

3.5 Creating a New Point Layer in a GeoPackage

Now create a new vector point layer by following the same steps.

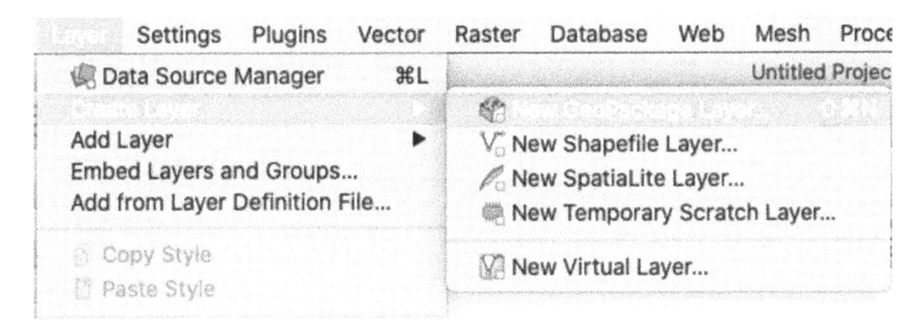

This time, choose *MultiPoint* as the *Geometry type*, and name the layer `ArchSites` for representing archaeology sites. Note the large number of types of vector files that can be created:

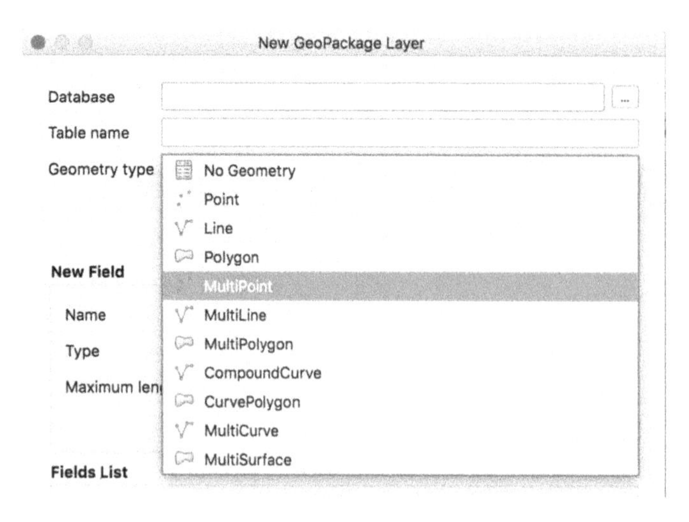

Use the *Add to Fields List* button to add a name (text), elevation (integer), and date (date) attribute.

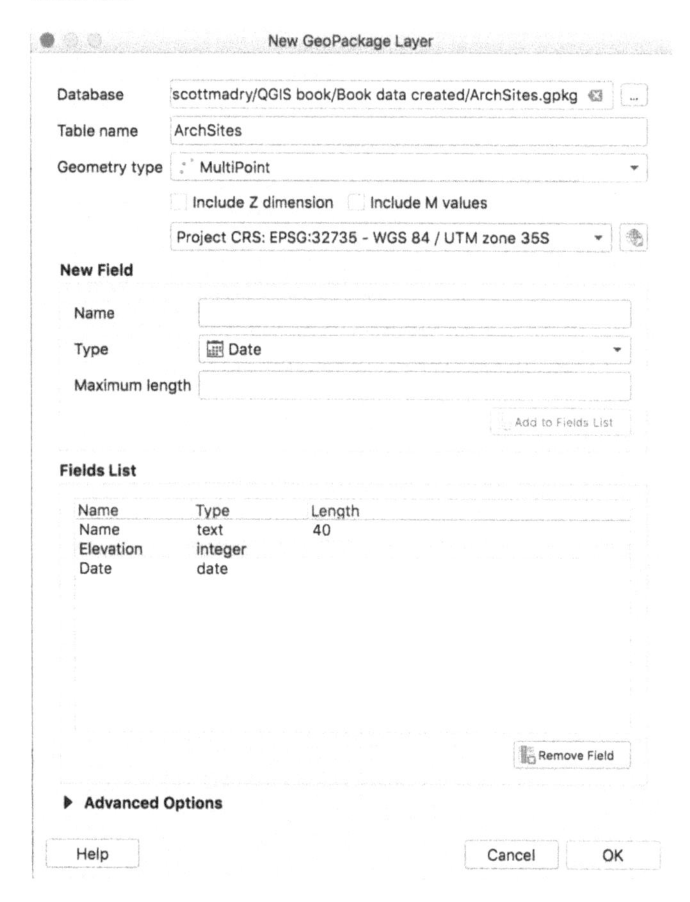

Place your points on the point elevation marks in the area near the walking trail. Call them 1, 2, 3, etc. Do however many you wish, and then click on the Toggle Editing icon and save your edits. You have just created a GeoPackage point layer.

Now right-click on the *ArchSites* name in the *Layers* panel, and click on `Open Attribute Table` to view the attributes of the layer you just created. Look at the bottom and you can see various options. Click on the pen to toggle the editing mode, then click on the *New Field* icon to add new fields and edit its contents. Be sure to set the width to enough spaces to type the name, date, and elevation.

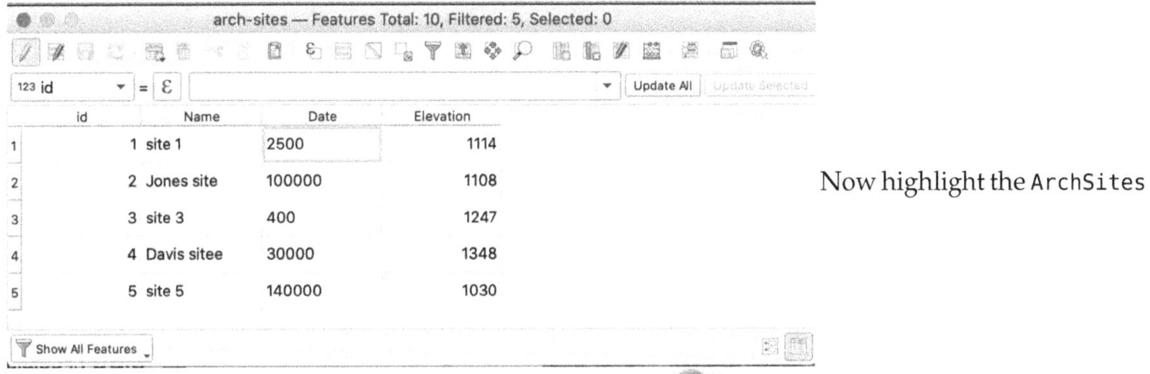

Now highlight the `ArchSites` in the *Layers* panel. You can use the *Identify Features* tool and click on any site to see the information about it. To see the advanced digitizing tools, click on a vector file in the *Layers* panel and right-click to `Toggle Editing`. Click on the `Edit` menu at top and you can see all the advanced editing features. Save your project, as you have added new data layers.

There is a lot more to good digitizing, and remember the GIS adage of "Garbage in - Garbage out!", so review the QGIS documentation and tutorials for more information and learn about good data creation skills. Always complete the metadata for newly created files, including who did it, when, where the data are stored, what the data were derived from, any problems, etc. Always conduct a quality control process, preferably having someone else review your data before it is cleared for use. And always back up your data, including maintaining an off-site backup.

4. Finding GIS Data Online

Getting started in GIS requires accurate and appropriate data for your study area. This may vary from the global to the most local scale, or even under the ocean or on the Moon or Mars. Access to accurate and current GIS vector and raster data varies widely by nation. Here are several excellent sources of free GIS and imagery data on the internet.

4.1 The Geofabric OSM Daily Shapefile Update

The Geofabric GmbH of Germany operates a website that creates a daily update of the OpenStreetMap (OSM) project. The update is provided by nation in shapefile format. The OpenStreetMap is a wiki mapping project, where people map their own neighborhoods in a crowdsourcing process. The OSM data are all available at the OSM website: `https://www.openstreetmap.org/` but these data are in the OSM's own format, and they are not readily useful in your GIS. QGIS does have a plugin to convert the data, discussed below, but the Geofabric site is a better way. This site provides a simple, accurate, and updated source of national scale GIS data that are free to download and use without commercial restrictions at: `http://download.geofabrik.de`

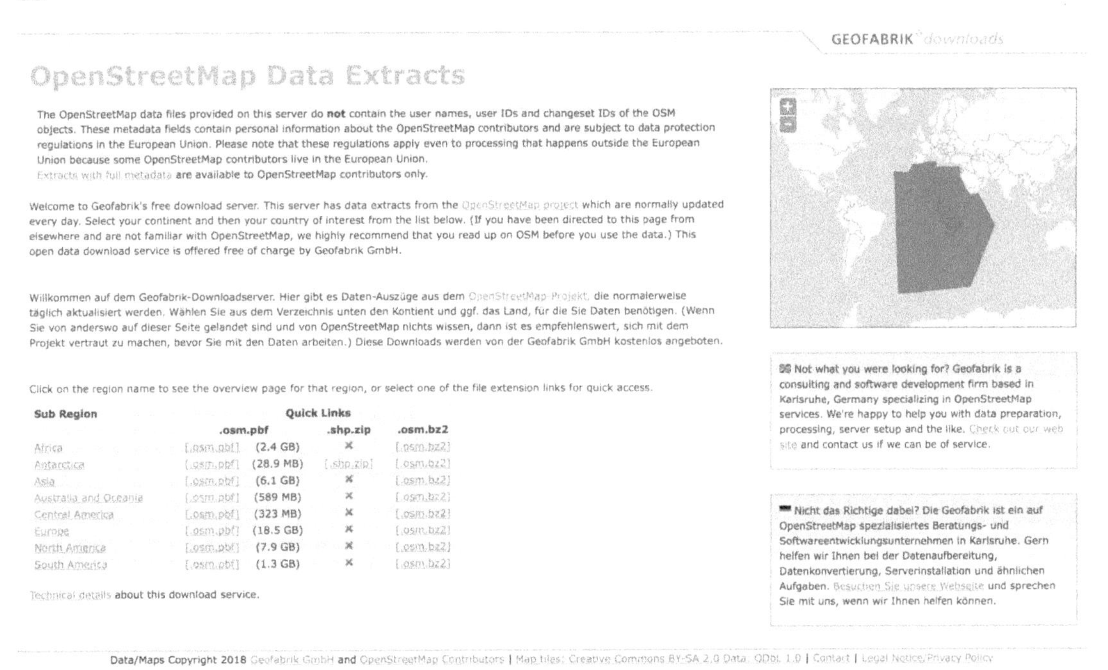

Click on the subregion or continent, and then the individual nation. Be sure to download the shapefiles option, .shp.zip (in the middle), and then unpack the .zip file and you will find that day's OpenStreetMap data available to you to use in QGIS, or any other GIS.

Sub Regions

Click on the region name to see the overview page for that region, or select one of the file extension links for quick access.

Sub Region	Quick Links		
	.osm.pbf	.shp.zip	.osm.bz2
Algeria	[.osm.pbf] (70 MB)	[.shp.zip]	[.osm.bz2]
Angola	[.osm.pbf](30.7 MB)	[.shp.zip]	[.osm.bz2]
Benin	[.osm.pbf](25.6 MB)	[.shp.zip]	[.osm.bz2]
Botswana	[.osm.pbf](35.4 MB)	[.shp.zip]	[.osm.bz2]
Burkina Faso	[.osm.pbf](35.5 MB)	[.shp.zip]	[.osm.bz2]
Burundi	[.osm.pbf](10.7 MB)	[.shp.zip]	[.osm.bz2]
Cameroon	[.osm.pbf] (121 MB)	[.shp.zip]	[.osm.bz2]

The South Africa download is 230 Mb, made up of 18 individual shapefiles, and also contains this ReadMe file:

> The files in this archive have been created from OpenStreetMap data and are licensed under the Open Database 1.0 License. See www.openstreetmap.org for details about the project.
>
> This file contains OpenStreetMap data as of 2018-09-04T20:14:02Z. Every day a new version of this file will be made available at:
> http://download.geofabrik.de/africa/south-africa-latest-free.shp.zip
>
> A documentation of the layers in this shape file is available here:
> http://download.geofabrik.de/osm-data-in-gis-formats-free.pdf
>
> Geofabrik also makes extended shapefiles to order; please see: http://www.geofabrik.de/data/shapefiles.html for details and example downloads.

Drag the shapefiles onto your QGIS map space, or use the other methods previously discussed to add them. This is a wonderful service to the international GIS community.

4.2 Natural Earth Data Global GIS data website

There is another excellent website with global scale GIS data called Natural Earth Data. This site provides public domain, global scale GIS data, including both vector and raster data such as shaded relief maps. The data come in several scales, ranging from 1:10 million to 1:100 million scales: http://www.naturalearthdata.com/downloads/

4.3 USGS EarthExplorer

The EarthExplorer from the US Geological Survey provides access to a wide variety of mostly remote sensing data across the world: `https://earthexplorer.usgs.gov`

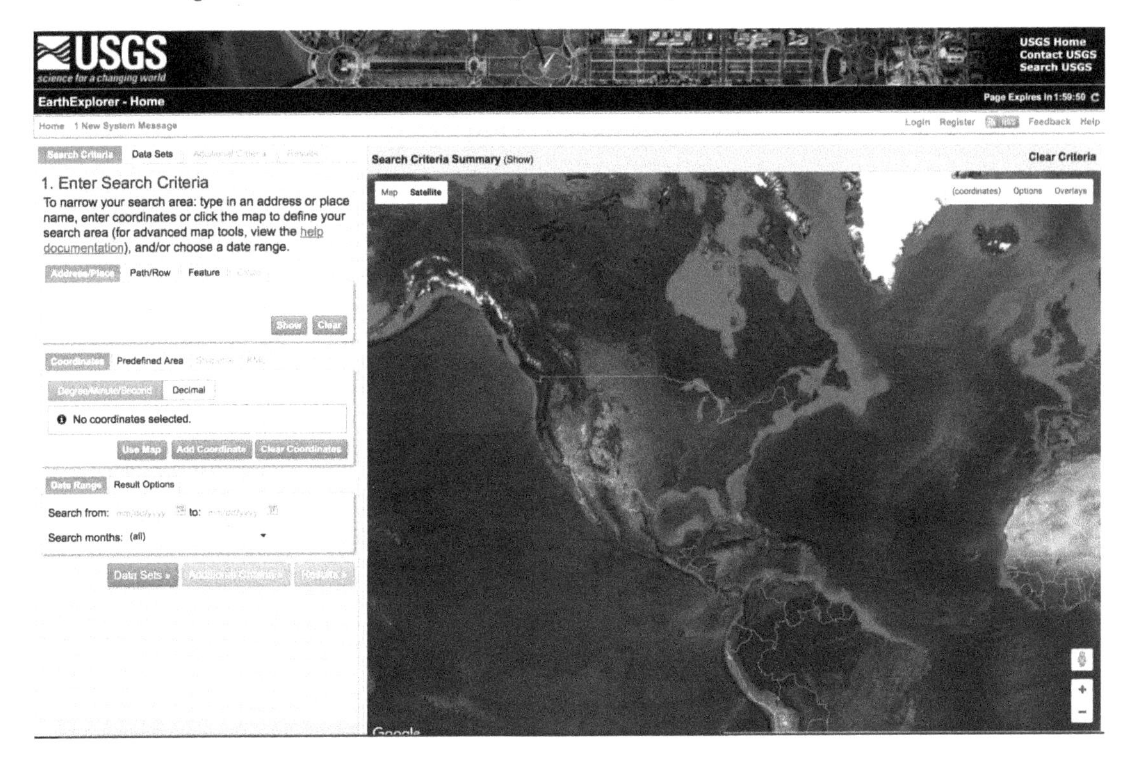

4.4 USGS National Map Portal

For data in the U.S., the U.S. Geological Survey also operates a variety of websites with free and comprehensive data for the USA at the National Map website: `https://nationalmap.gov`

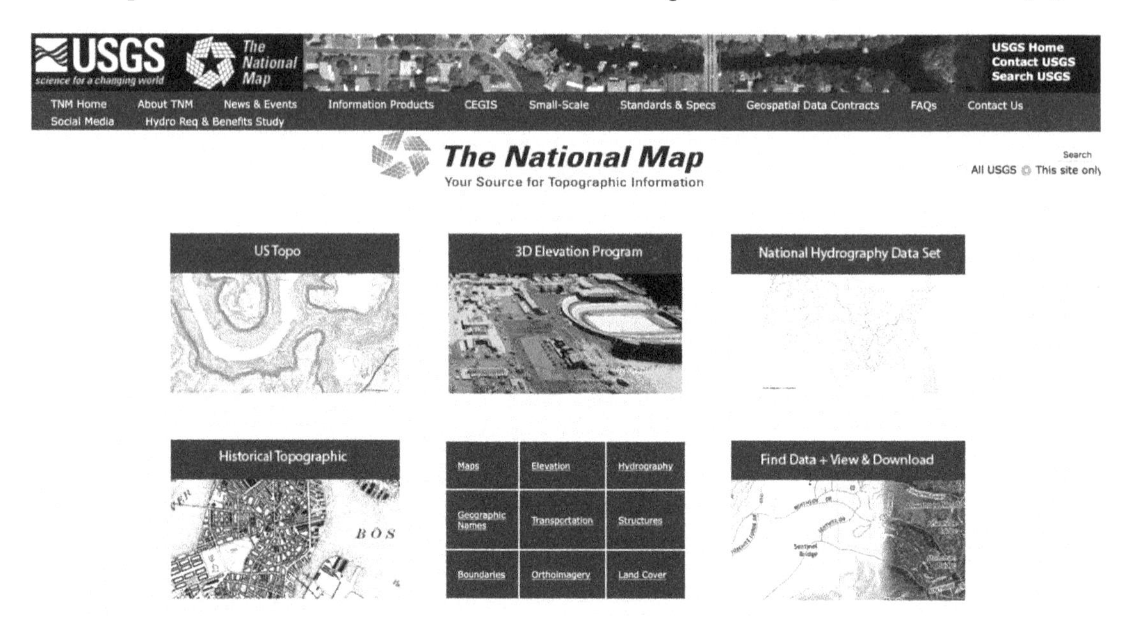

4.5 ESRI Open Data website

The ESRI Open Data website provides a portal for accessing a wide variety of GIS data sets from `https://hub.arcgis.com/pages/open-data`. Be aware that this is a subscription only website, and you have to have an ArcGIS Online account to access it.

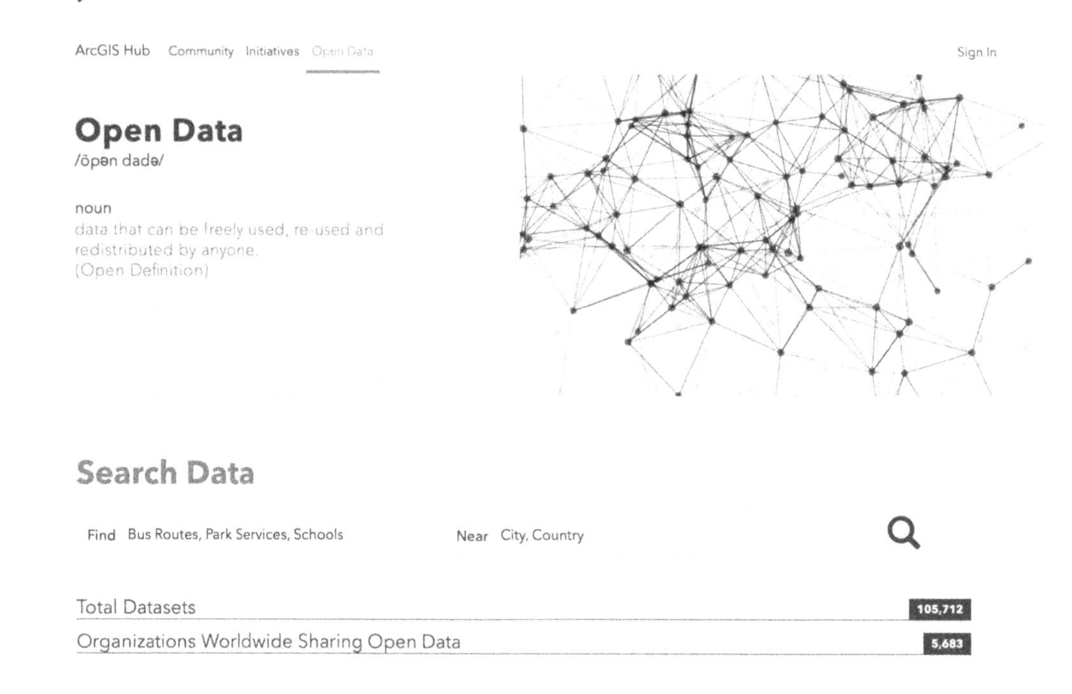

4.6 DIVA GIS website

This site provides a source of national scale GIS data from various sources. You can download by country and individual vector or raster data type: `http://www.diva-gis.org/gdata`. Much of the data is less detailed or updated than the Geofabrik site discussed above, but it also contains useful raster and demographic data.

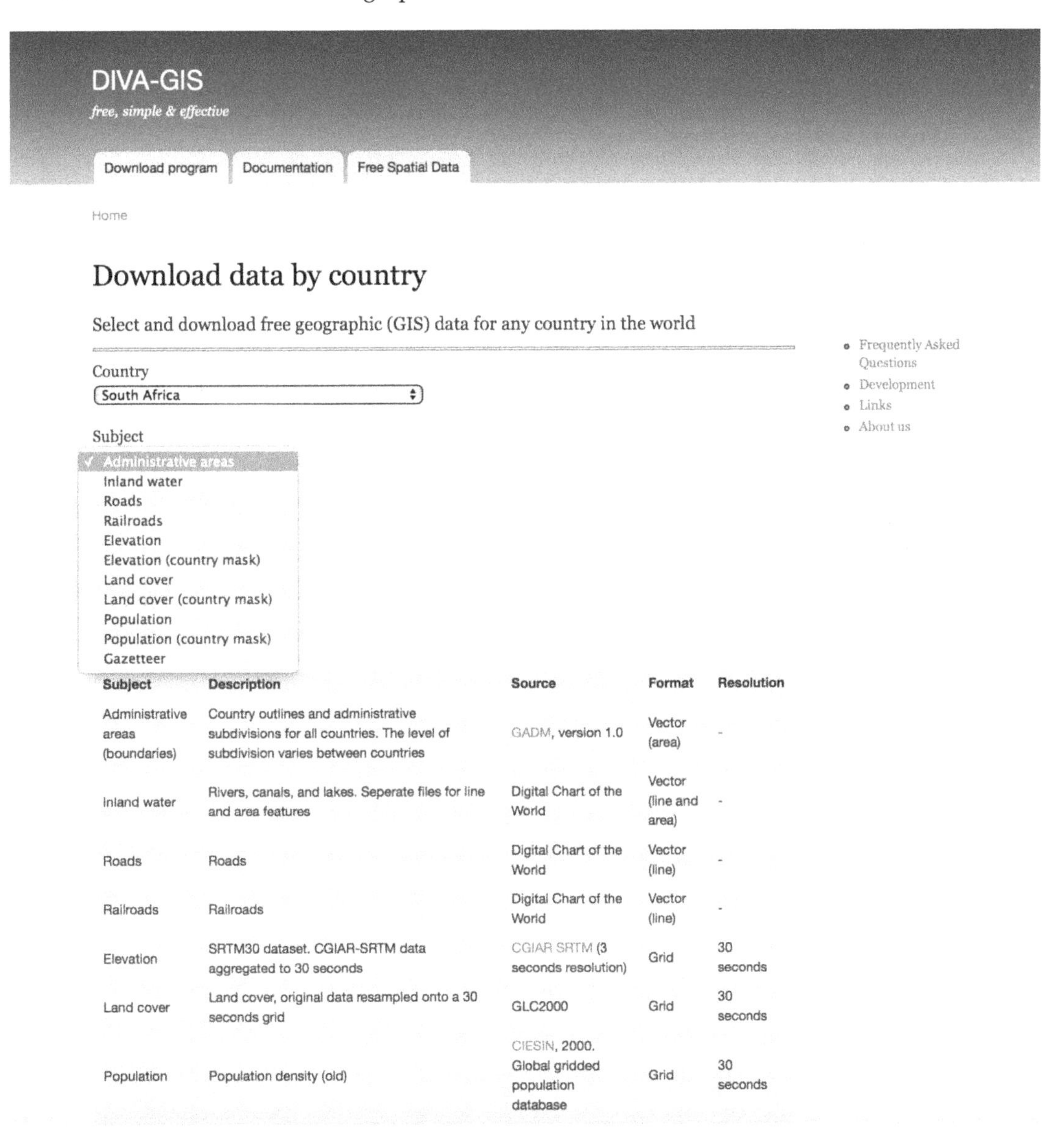

Subject	Description	Source	Format	Resolution
Administrative areas (boundaries)	Country outlines and administrative subdivisions for all countries. The level of subdivision varies between countries	GADM, version 1.0	Vector (area)	-
Inland water	Rivers, canals, and lakes. Seperate files for line and area features	Digital Chart of the World	Vector (line and area)	-
Roads	Roads	Digital Chart of the World	Vector (line)	-
Railroads	Railroads	Digital Chart of the World	Vector (line)	-
Elevation	SRTM30 dataset. CGIAR-SRTM data aggregated to 30 seconds	CGIAR SRTM (3 seconds resolution)	Grid	30 seconds
Land cover	Land cover, original data resampled onto a 30 seconds grid	GLC2000	Grid	30 seconds
Population	Population density (old)	CIESIN, 2000. Global gridded population database	Grid	30 seconds

4.7 The OpenTopography website

The OpenTopography website provides access to high resolution DEM data, including data from LiDAR, around the world. The program is located at the San Diego Supercomputer Center at the University of California, San Diego: `https://opentopography.org`

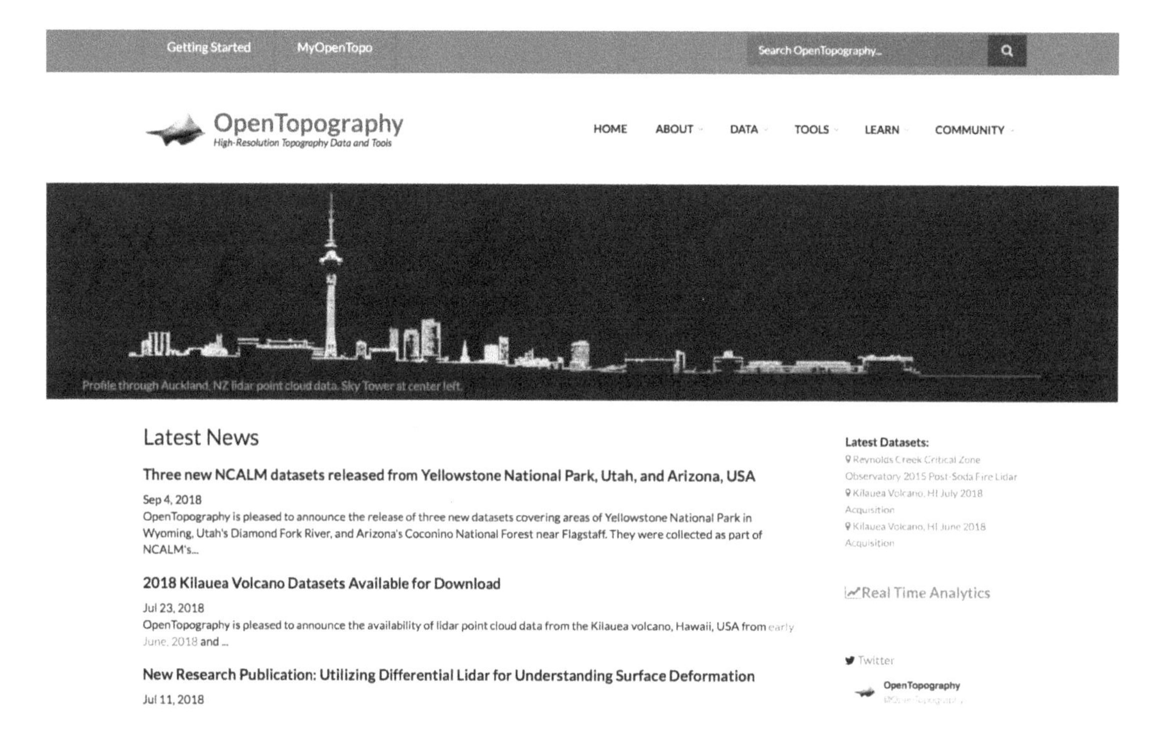

4.8 Other Online GIS Data Sources

Many of the states and major cities in the U.S. also provide online access to their GIS data, like the California Geoportal: `https://gis.data.ca.gov` or the Charlotte, NC Open Mapping site: `http://maps.co.mecklenburg.nc.us/openmapping/index.html`

Additionally, your local university library very likely has a GIS librarian and a website that provides pointers to accessing GIS data. Excellent examples include the Pennsylvania Geospatial Data Clearinghouse `https://www.pasda.psu.edu`, the Stanford University libraries Spatial Data Resources page: `https://library.stanford.edu/research/stanford-geospatial-center/data`, and the NC State University libraries GIS download page: `https://www.lib.ncsu.edu/gis/dataweb`.

Many national governments provide geospatial data, as well as international agencies. There is no single GIS data repository, so you will have to search in your local area on the web. Quality GIS data is vital to useful analysis. Please consider carefully the source, date, scale, and other factors of all the data you use. This is why it is so important to do your metadata so we can track the lineage of GIS data. Many agencies require fully compliant metadata to be created under funded projects.

5. Vector Data Import and Export

5.1 Vector data import and export

While QGIS often uses shapefiles and GeoPackage vector files, it is simple to both import and export vector data of a wide variety of formats using QGIS. Click on `Layer`, and go to `Add Layer`, and you will see the many types of data that can be imported.

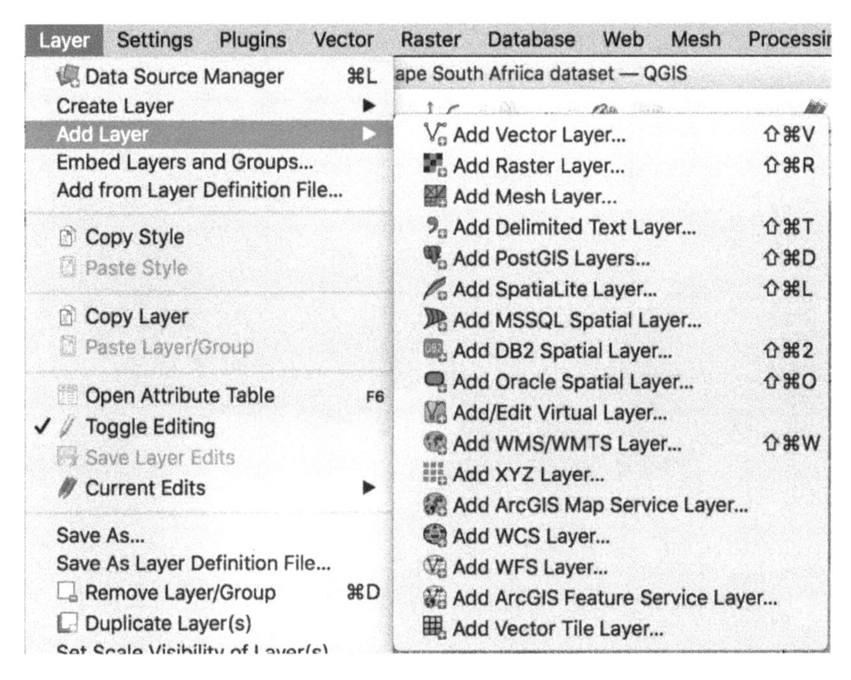

Click on `Add Vector Layer` at the top, and it will open the *Data Source Manager* dialog:

This shows you all the data types that you can add into QGIS, with all of the data types shown along the left-hand edge.

Click on the three dots icon ... under *Source*, at right and you can browse to the location of your file.

While browsing for a file, click the drop-down box that initially contains *All files* to see the many formats of vector files that can be accessed:

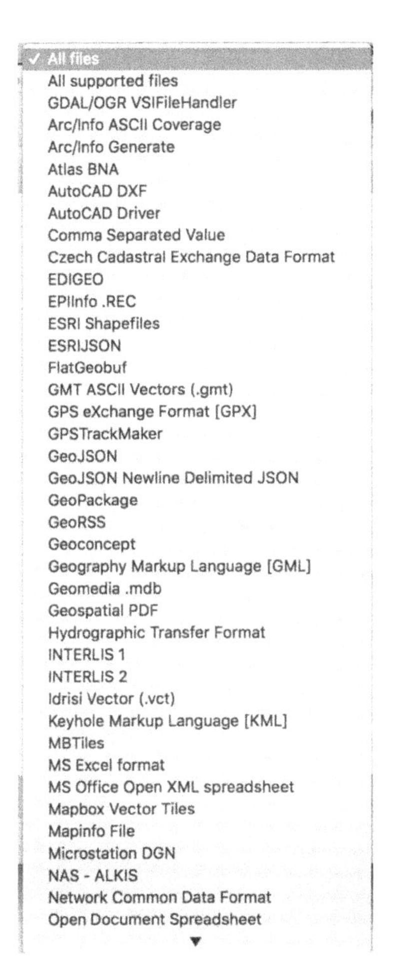

This is only half of the many different vector data types we can use in QGIS. There are three other options along the top of the *Data Source Manager - Vector* dialog that are for other commonly used data sources.

Under *Directory* you have the following options:

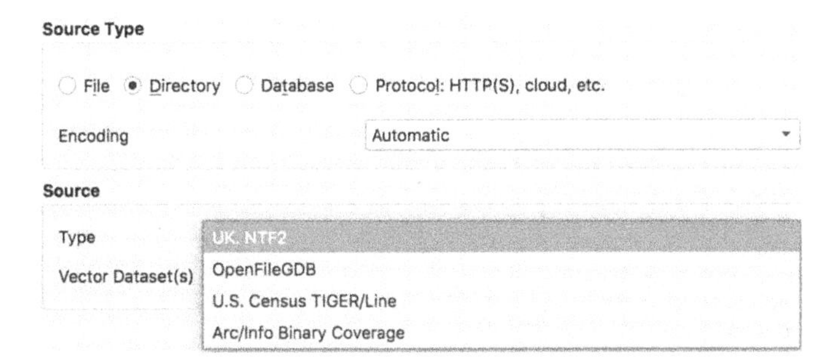

UK NTF2 files are a U.K. Ordnance Survey digital terrain format, supplied in NTF version 2 format.

OpenFileGDB gives access to files using the ESRI File Geodatabase (.gdb) format. You can open ArcGIS geodatabases and QGIS will separate the individual files into shapefiles after you indicate which you want. OpenFileGDB can read ArcGIS 9.x Geodatabases and up. For info on ESRI Geodatabase see: `http://desktop.arcgis.com/en/arcmap/10.3/manage-data/administer-file-gdbs/file-geodatabases.htm`

The U.S. Census Bureau produces vector TIGER files as a part of the ten year census process. This includes census blocks and other locations that can be used with census demographic data, as well as vector roads, streams, etc. For more info, see: `https://www.census.gov/programs-surveys/geography.html` They also provide data as shapefiles: `https://www.census.gov/geographies/mapping-files/time-series/geo/tiger-line-file.html`

Arc/Info Binary Coverages are older vector data from ESRI's Arc/Info V7 and earlier).

Under *Database*, you can access ODBC, ESRI personal GeoDatabases, MSSQL, and PostgreSQL databases.

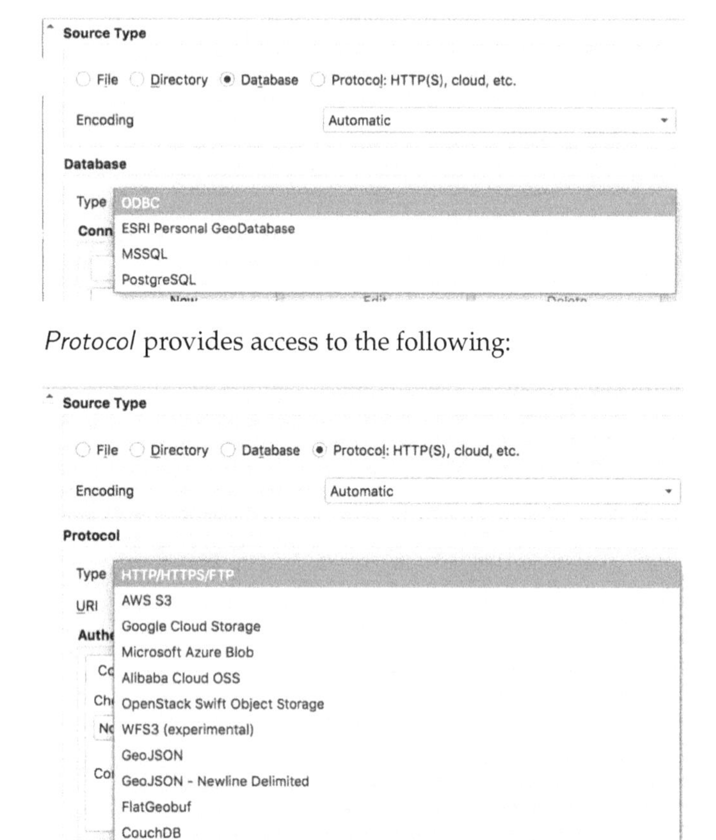

Protocol provides access to the following:

HTTP/HTTPS/FTP files are well known. AWS S3 is the Amazon cloud computing web service. This provides data storage through a web services interface. You can also access Google Cloud Storage data here, MS Azure, and Alibaba cloud OSS, OpenStack, etc. GeoJSON is an important and increasingly popular open standard for representing geographical features, as well as their attributes. It is based on JavaScript Object Notation. To learn more, visit `http://geojson.org` and `http://www.macwright.org/2015/03/23/geojson-second-bite.html` FlatGeoBuf is a performant binary encoding for geographic data, based on Flatbuffers: `https://gdal.org/drivers/vector/flatgeobuf.html` Apache CouchDB is a database structure that uses JSON for documents, JavaScript for indexes, and HTTP as its API. More information at: `http://couchdb.apache.org`.

Converting File Formats

If you want to convert a shapefile or other vector file to another format, use the `LAYER > SAVE AS` menu to open the *Save Vector Layer as* dialog:

Click on the *Format* drop-down box and you will see the many formats available:

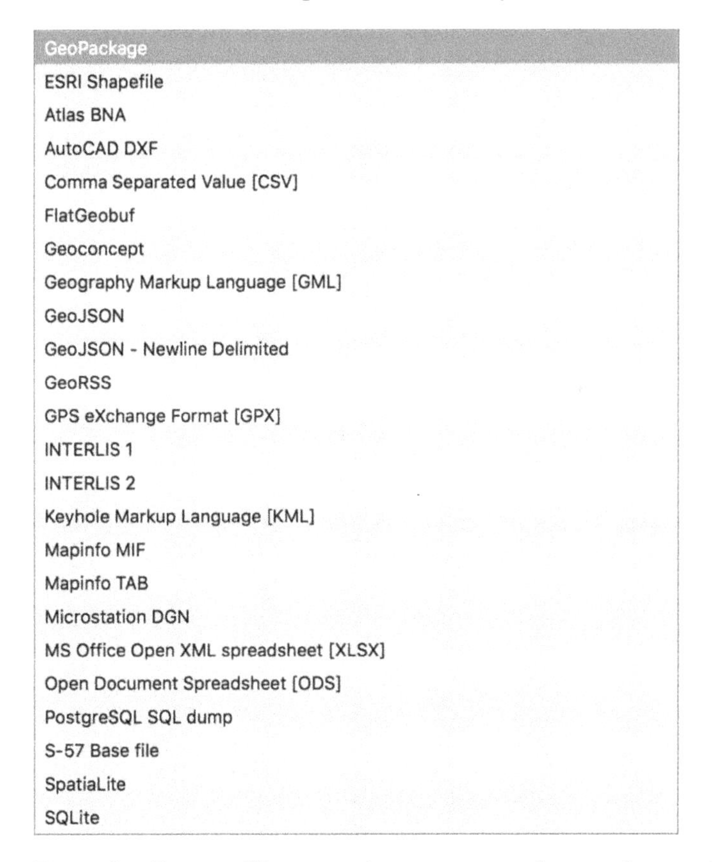

Enter the *Format*, *File name*, *Layer name*, and any other options needed and click OK to create the new file. There is a comparable process for raster data that will be covered later.

5.2 Open Geospatial Consortium (OGC) Data

You can access a variety of OGC format data from the *Data Source Manager*. Click on `Layer->Data Source Manager` to see:

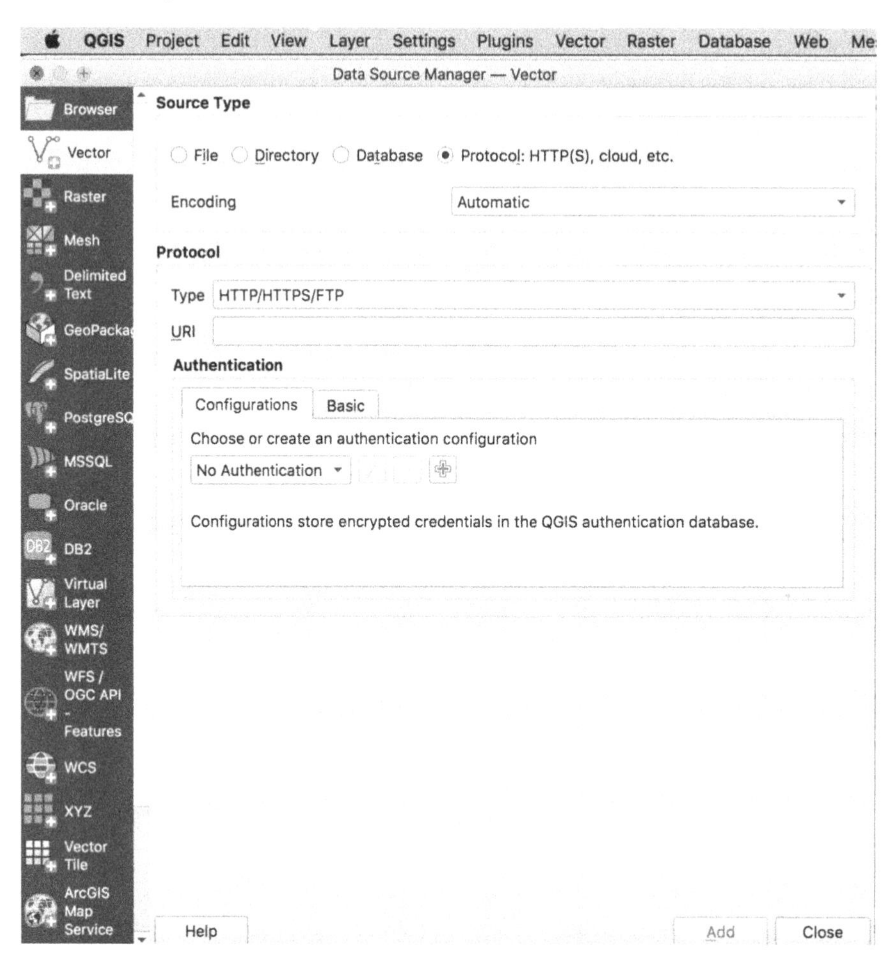

Click on the icons along the left side to see the various options.

The OGC (`http://opengeospatial.org`) is an international, non-profit consortium of government agencies, universities, and private companies that have agreed upon a wide range of spatial data formats and protocols (I was a founding board member way back in the early 1990's). The current list of OGC data that can be used by QGIS includes:

- WMS — Web Map Service (WMS/WMTS Client)
- WMTS — Web Map Tile Service (WMS/WMTS Client)
- WFS — Web Feature Service (WFS and WFS-T Client)
- WFS-T — Web Feature Service - Transactional (WFS and WFS-T Client)
- WCS — Web Coverage Service (WCS Client)

- SFS — Simple Features for SQL (PostGIS Layers)
- GML — Geography Markup Language

WMS (Web Mapping Service) is a standard protocol for serving georeferenced data over the Internet, along with WMTS (Web Mapping Tile Service) data. The main difference is that WMS serves an image that cannot be edited, while WMTS serves files that are georeferenced tiles. Also available are WFS (Web Feature Service), which are fully editable in your GIS, and WCS (Web Coverage Service), all standard data formats of OGC. To learn more about OGC, see `http://opengeospatial.org`, and for the complete QGIS OGC manual, see: `http://docs.qgis.org/2.8/en/docs/user_manual/working_with_ogc/ogc_client_support.html`

Choose `Layer->Add Layer->Add WMS/WMTS Layer` from the menu to open the *Data Source Manager* dialog for WMS/WMTS:

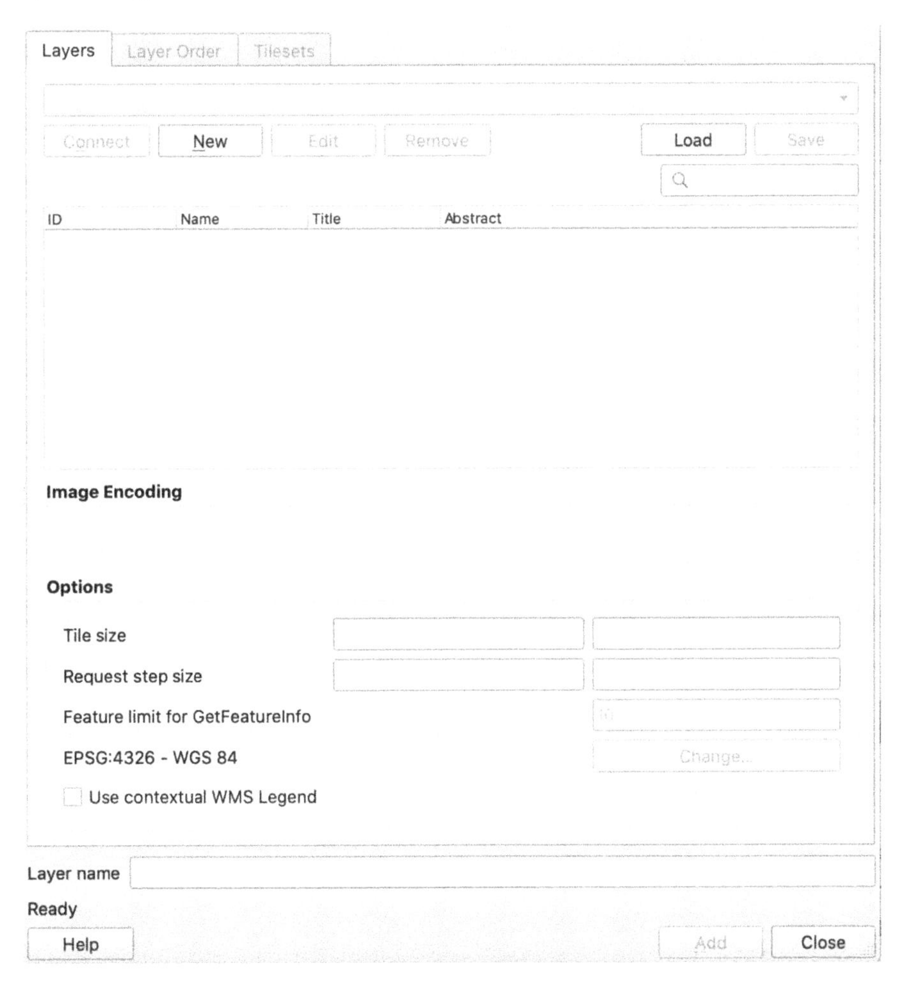

Click *New* at the top and for *Name* enter *Mapserver Demo* and for *URL*, `http://demo.mapserver.org/cgi-bin/wms`, then click *OK* to add the server:

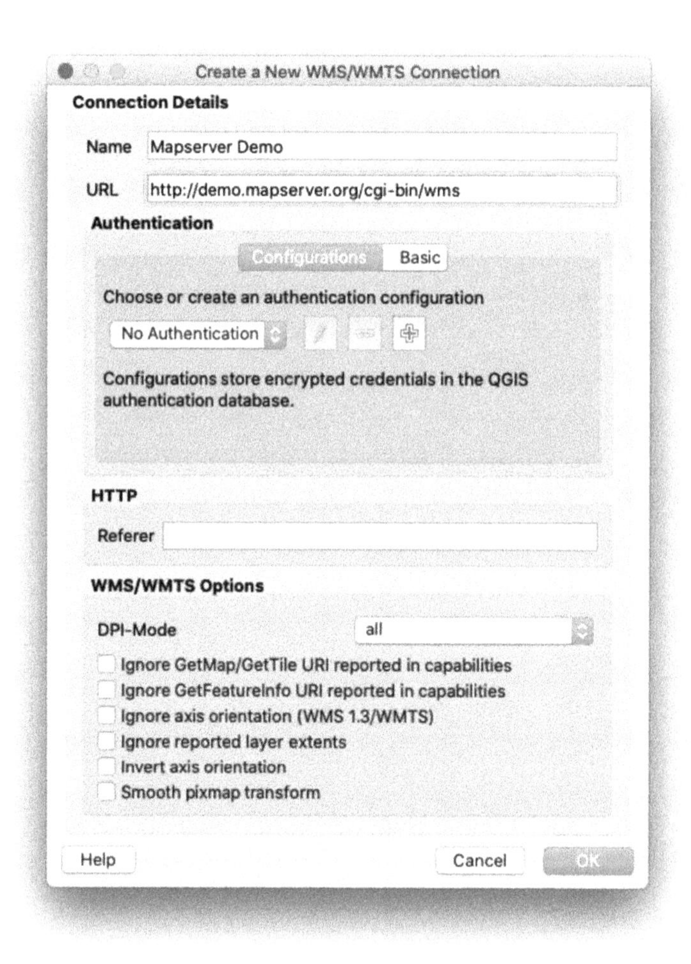

You'll now see the new server available. Click *Connect* and the four layers available will be shown in the list.

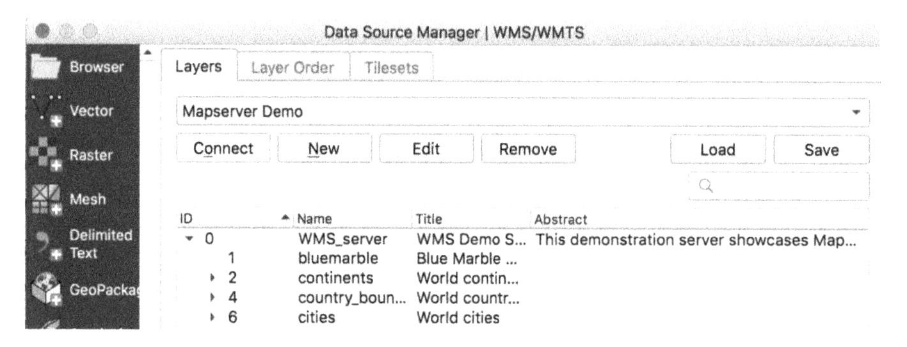

Click on the *Blue Marble* dataset, then click *Add*.

You may need to set your web proxy settings, as shown below. Clicking *Add* will add the layer to your *Layers* panel. Right-click on the newly added layer and choose `Zoom to Layer` to see the global raster satellite image. You can zoom in South Africa.

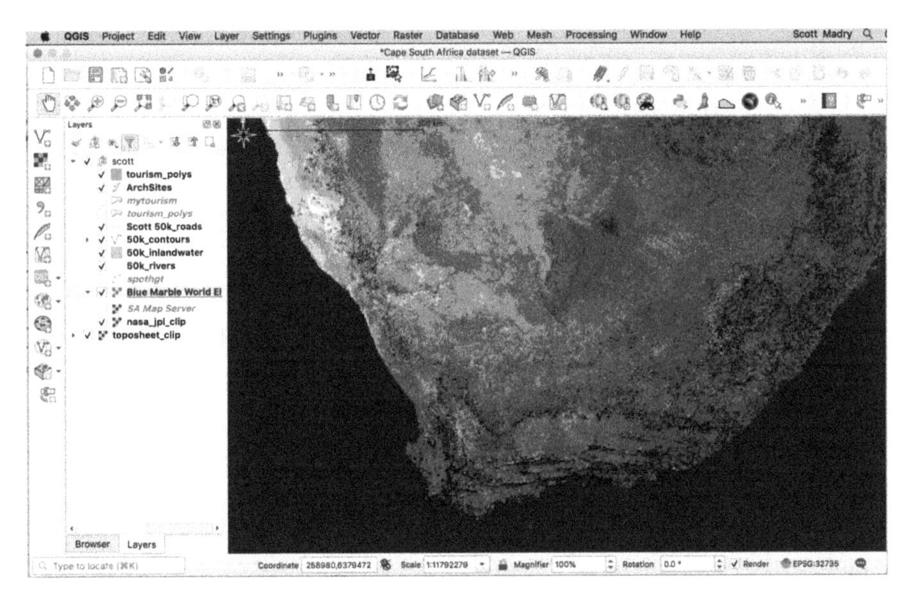

Go ahead and remove the image from your *Layers* panel by right-clicking on it and choosing `Remove Layer`. For an up to date list of WMS and WFS servers see: `http://www.skylab-mobilesystems.com/en/wms_serverlist.html`

WFS servers are much the same, but instead of just images or data you can see but not alter, they serve files that behave like any other QGIS file, where you can see the attributes, save, alter, etc. In QGIS, a WFS layer behaves pretty much like any other vector layer. You can identify and select features and view the attribute table. Editing requires WFS-T.

Adding a WFS layer is very similar to the procedure used with WMS. The difference is there are no default servers defined, so we have to add our own.

You can see under `Layer->Add Layer`

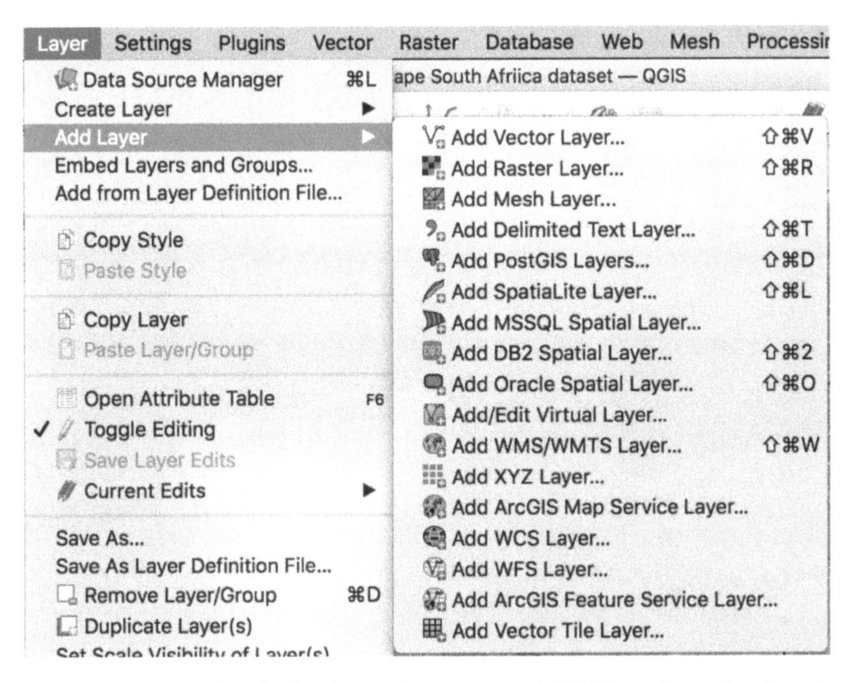

that there are multiple database, imagery, and GIS data download options, including WCS and

WFS (web coverage services and web feature services).

There are many other sources and formats of GIS data, and the good news is that QGIS can import, use, and export the vast majority of them.

6. Vector Data Processing and Analysis

6.1 The Vector Menu

Once we download or create our vector data, we will want to do some analysis and further manipulation. Let's look at some of the vector processing functions in QGIS. Click on the `Vector` menu.

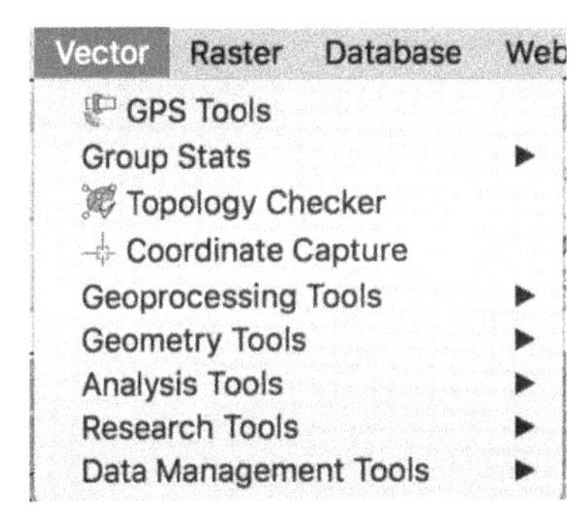

You will see several groups of modules, and these work on any vector layer, regardless of provider. The GPS tools at top will be discussed later.

6.2 Vector Geoprocessing Tools

QGIS has full vector geoprocessing abilities. Under the `Vector->Geoprocessing Tools` menu you'll see the standard "McHargian" Boolean analysis functions: buffer, union, intersection, clip, etc.

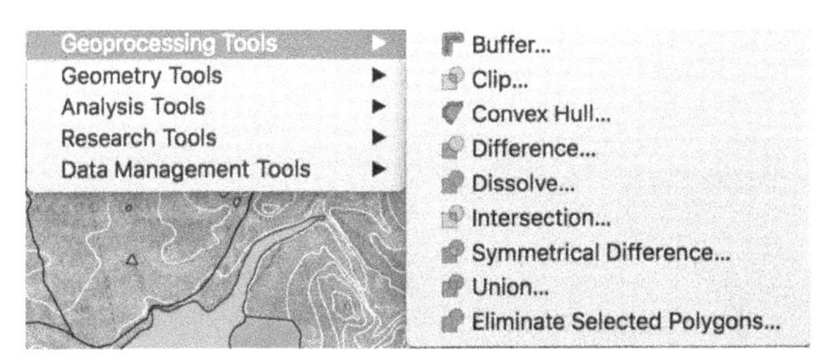

If you are not aware of Ian McHarg and his impact on GIS, you should read about his life (`https://en.wikipedia.org/wiki/Ian_McHarg`). To learn about Ian McHarg's ground-breaking early use spatial analysis and overlaying data in GIS, a fascinating bit of GIS history, see `http://www.ou.edu/class/webstudy/fehler/E3/go/introduction.html`. For further information on using Boolean logic in GIS, see: `http://www.gitta.info/Suitability/en/html/BoolOverlay_learningObject2.html`

6.3 Vector Buffers

Creating a vector buffer is simple in QGIS. Click on `Vector > Geoprocessing Tools > Buffer` and fill in the dialog box shown below, setting a 500 meter buffer for our *50k_ inlandwater* layer:

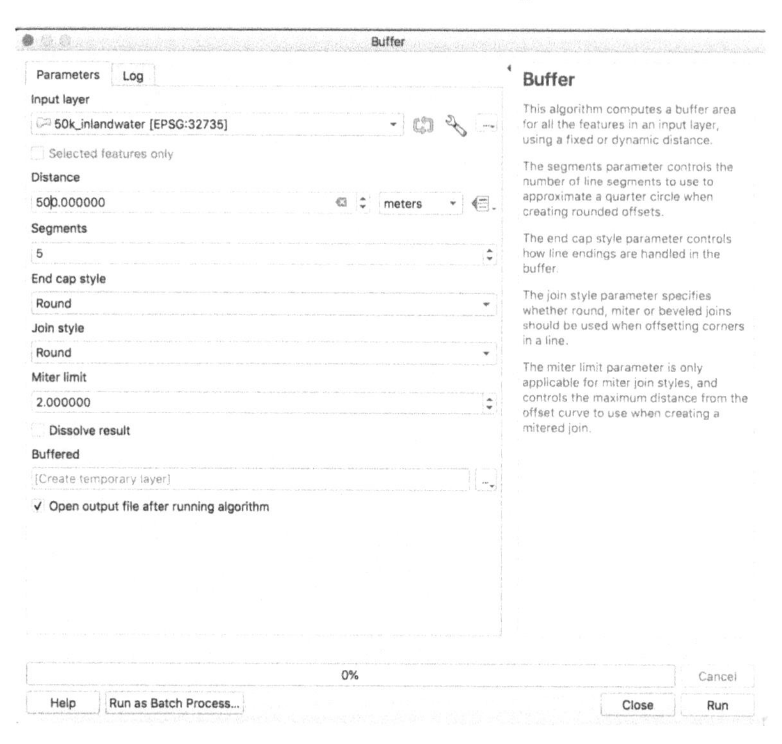

In order to use the geoprocessing tools, you have to load the layer into QGIS. Remember that this is now a metric database, in a UTM projection, so the distance will be 500 meters. Click on *Run* at bottom right, and the results look like this:

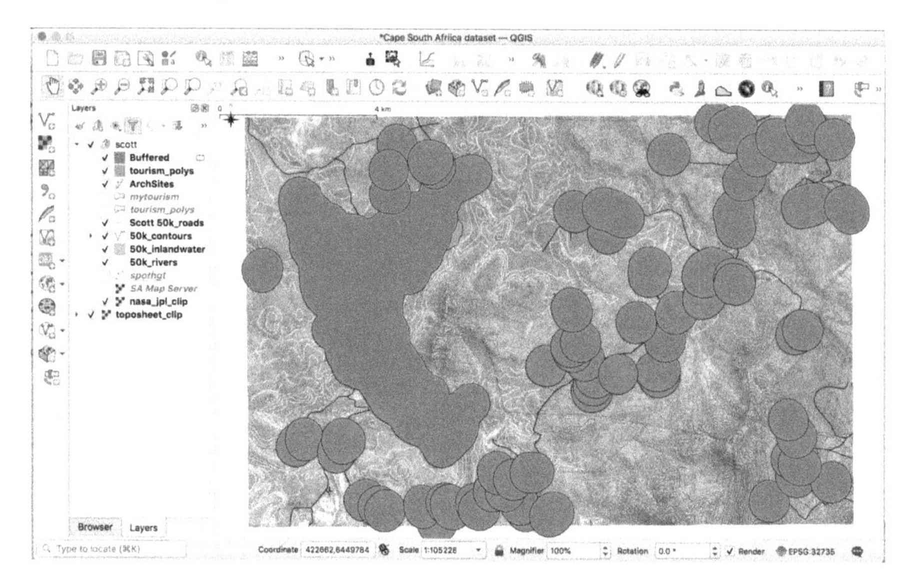

Here you see (in a random color, here a magenta) the 500 meter buffer created around each inland water feature. Place the buffer layer below the other vectors in the *Layers* panel at left so you can see it better.

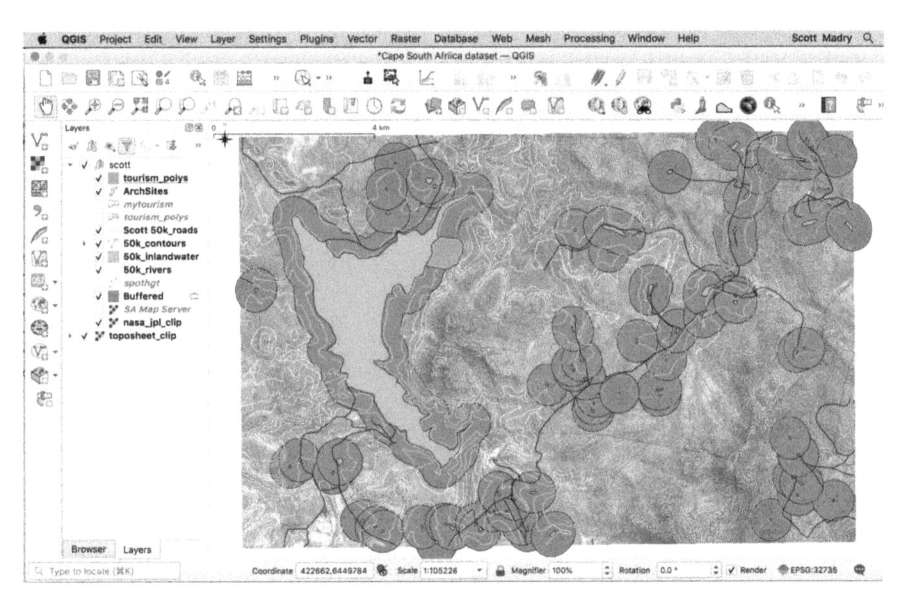

The *Log* tab on the *Buffer* dialog shows you what was done:

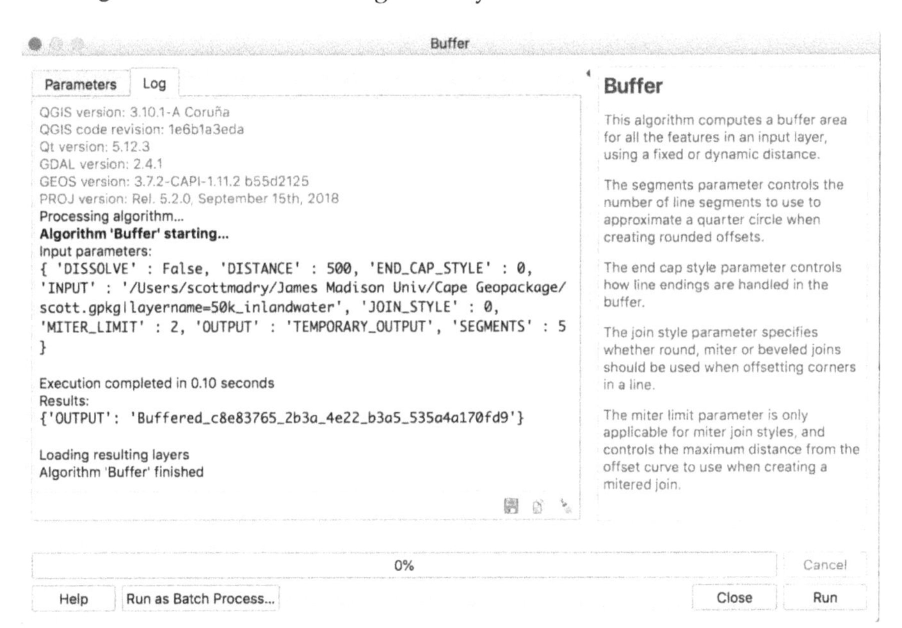

We have created a new vector layer that shows all areas that are within 500 meters of a lake or pond. Note that it will default to creating a temporary file, which is indicated with an icon to the right of it's name in the *Layers* panel:

The temporary buffer will be automatically deleted when you close your project, so you might want to save it as a permanent file for later use. To do this, make sure it is highlighted in the *Layers* panel, then click on the small icon to the right of the name to bring up the *Save Scratch Layer* dialog:

Note that it defaults to a GeoPackage format, but you have other options as well:

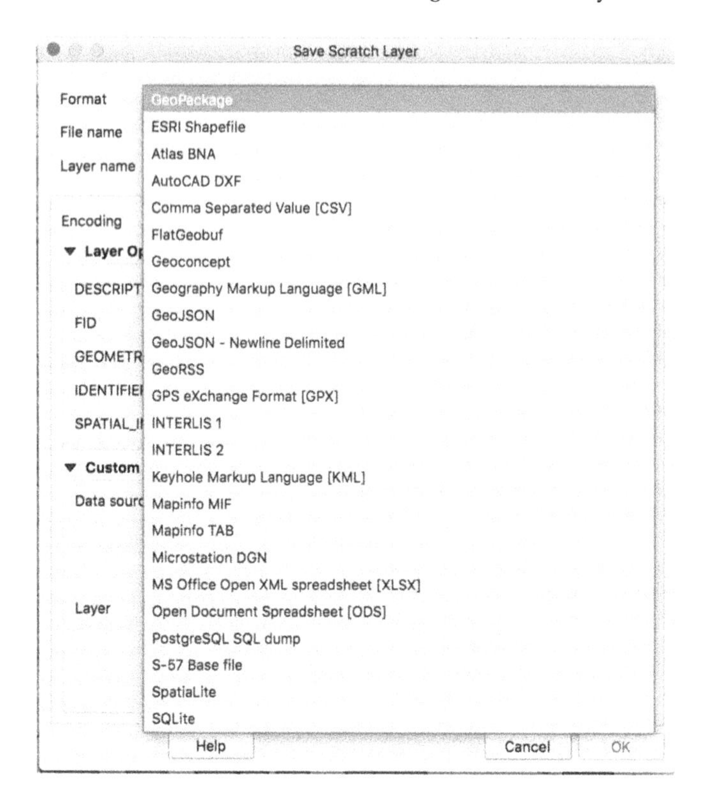

Now create a 500 meter buffer around the `spothgt` point file:

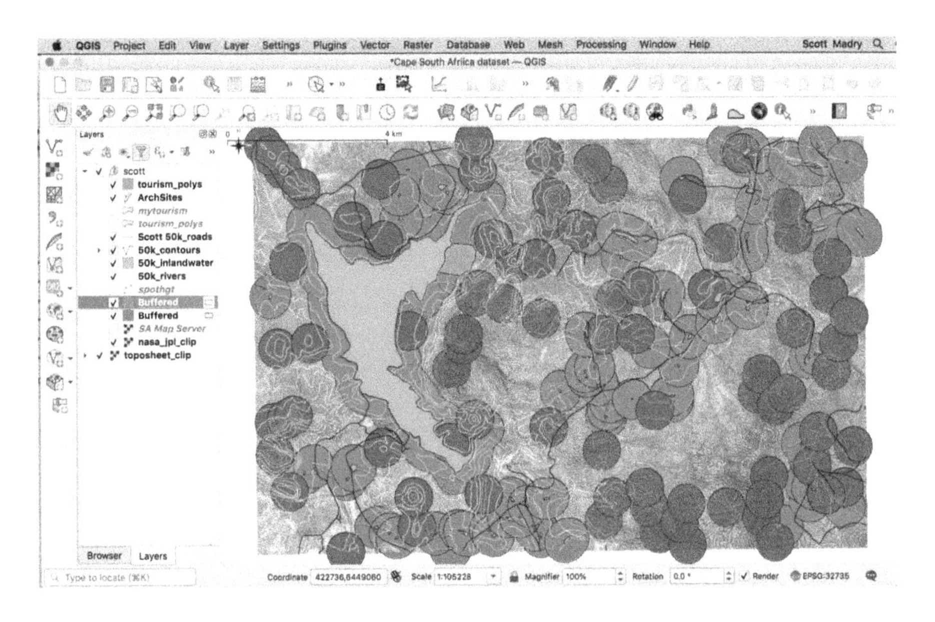

6.4 Vector Intersections

Now lets look at the intersection of the two buffers we just created. Click on `Geoprocessing Tools->Intersection` to bring up the *Intersection* dialog.

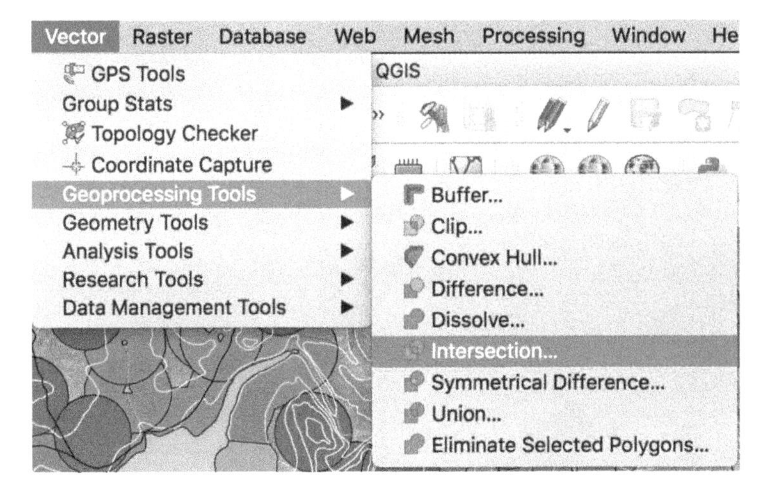

Choose the two new buffers layers we created:

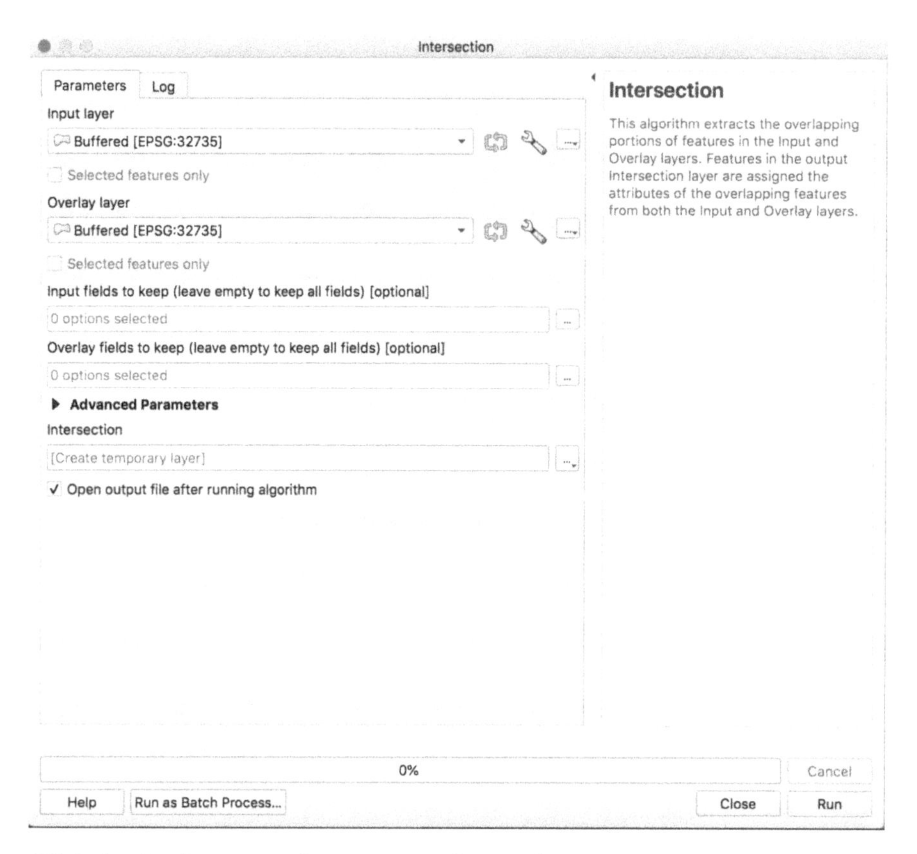

Click the *Run* button and we can see the result:

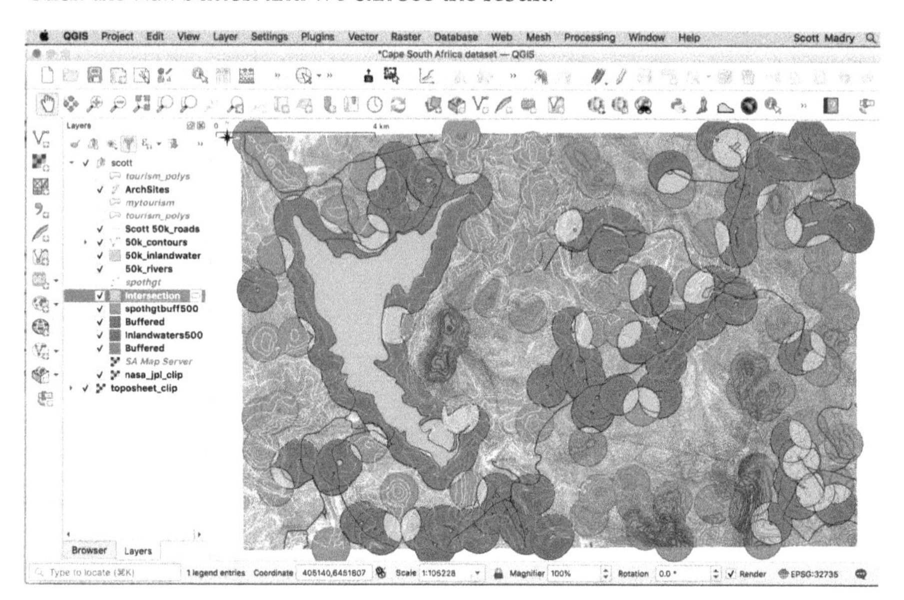

Now all the areas that are both within 500 meters of an inland water feature and our point file are displayed in a tan color. Turn off all the other layers and you will see just the intersecting polygons:

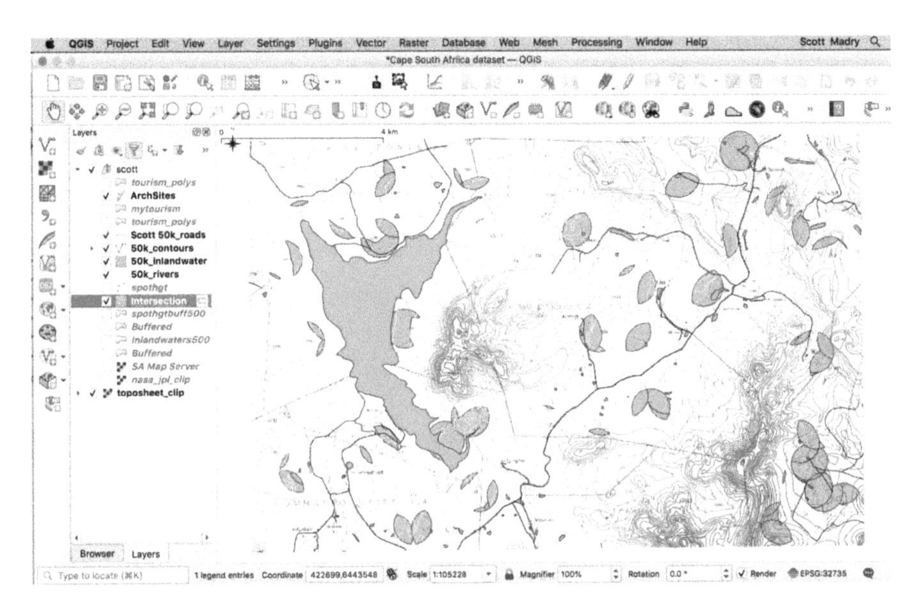

Next, look at the attribute table for the new `Intersection` layer:

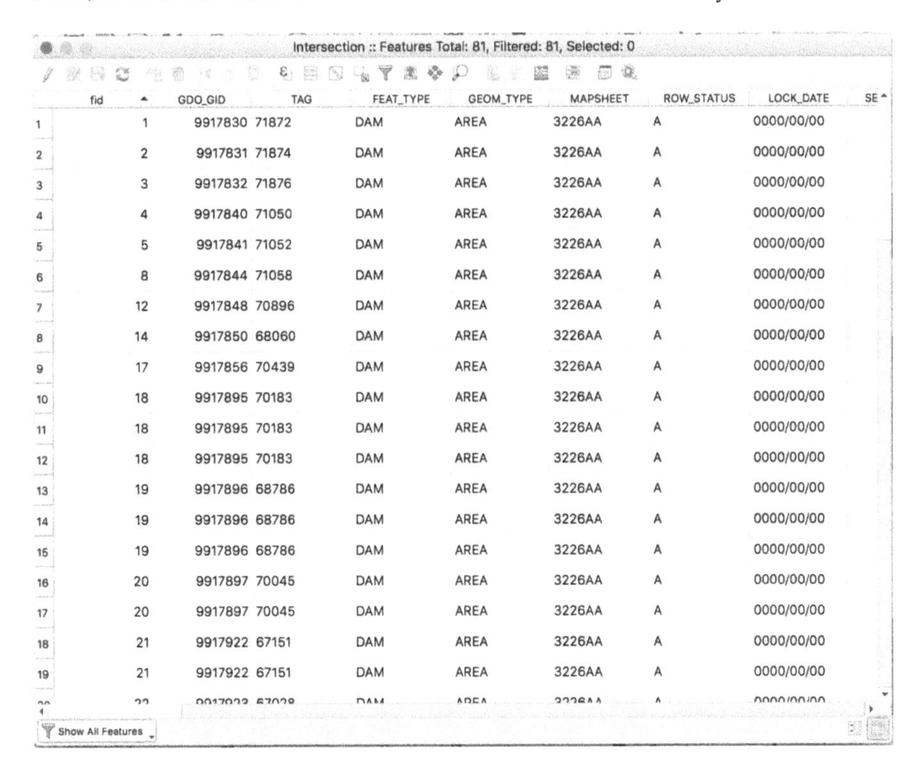

	fid	GDO_GID	TAG	FEAT_TYPE	GEOM_TYPE	MAPSHEET	ROW_STATUS	LOCK_DATE
1	1	9917830	71872	DAM	AREA	3226AA	A	0000/00/00
2	2	9917831	71874	DAM	AREA	3226AA	A	0000/00/00
3	3	9917832	71876	DAM	AREA	3226AA	A	0000/00/00
4	4	9917840	71050	DAM	AREA	3226AA	A	0000/00/00
5	5	9917841	71052	DAM	AREA	3226AA	A	0000/00/00
6	8	9917844	71058	DAM	AREA	3226AA	A	0000/00/00
7	12	9917848	70896	DAM	AREA	3226AA	A	0000/00/00
8	14	9917850	68060	DAM	AREA	3226AA	A	0000/00/00
9	17	9917856	70439	DAM	AREA	3226AA	A	0000/00/00
10	18	9917895	70183	DAM	AREA	3226AA	A	0000/00/00
11	18	9917895	70183	DAM	AREA	3226AA	A	0000/00/00
12	18	9917895	70183	DAM	AREA	3226AA	A	0000/00/00
13	19	9917896	68786	DAM	AREA	3226AA	A	0000/00/00
14	19	9917896	68786	DAM	AREA	3226AA	A	0000/00/00
15	19	9917896	68786	DAM	AREA	3226AA	A	0000/00/00
16	20	9917897	70045	DAM	AREA	3226AA	A	0000/00/00
17	20	9917897	70045	DAM	AREA	3226AA	A	0000/00/00
18	21	9917922	67151	DAM	AREA	3226AA	A	0000/00/00
19	21	9917922	67151	DAM	AREA	3226AA	A	0000/00/00

6.5 Vector Polygon Counts

If you look closely at the attribute table, you'll see that the intersection operation has combined the attributes of both source layers. Click on the *Sigma* ∑ statistical summary icon on the *Attributes* toolbar in the main QGIS window. For the `Intersection` layer, choose the *fid* category:

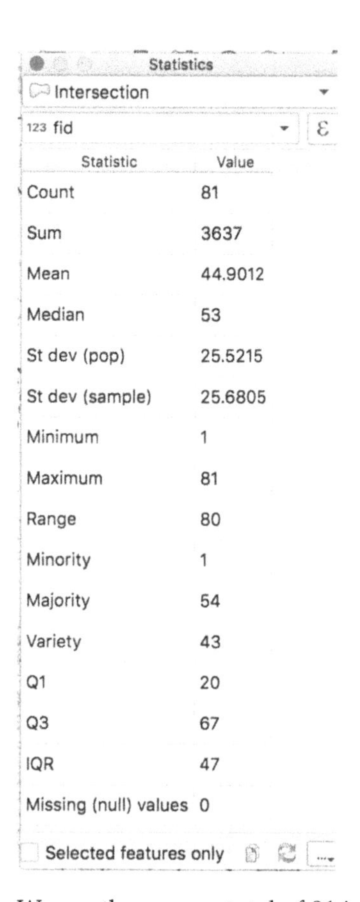

We see there are a total of 81 intersecting polygons. But how much total area is there?

6.6 Calculating the Area

To calculate the area of a vector layer, you should first open the attribute table and turn on editing. This will activate all of the icons at the top of the attribute table. Open the attribute table by right-clicking on the layer and choosing `Open Attribute Table`, then click on the *Field Calculator* icon.

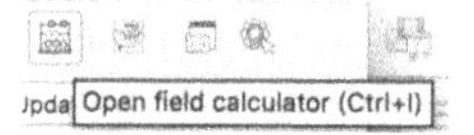

You will see the *Field Calculator* dialog:

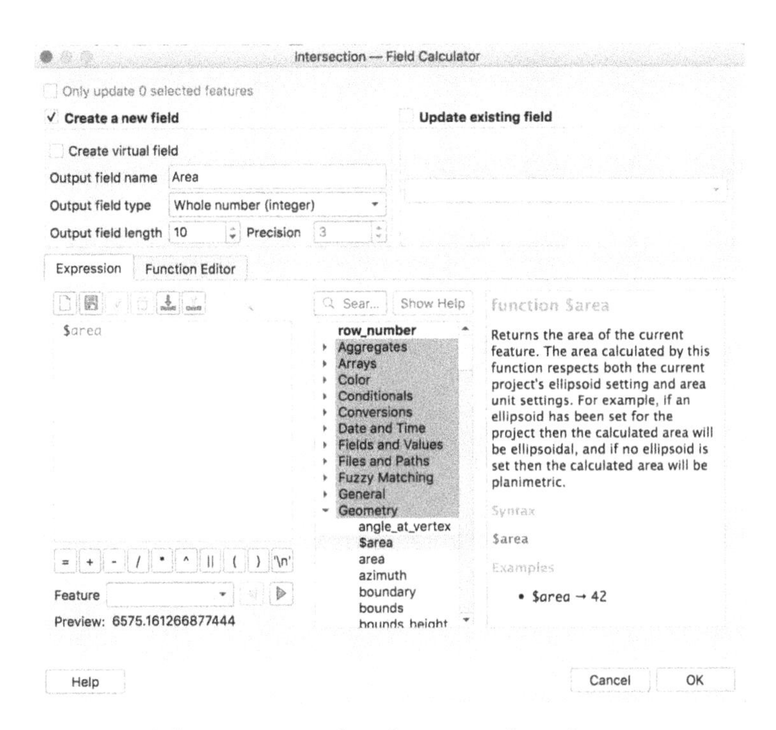

In the middle section, under *Geometry*, there is an operator `$area` that calculates the area of each row in the table. Double-click on *$area* and it will appear at left in the calculator window. Give the new field a name, as shown above, and click *OK*. You have now added a new attribute field with the area of each polygon, in square meters. Save the attribute file by clicking on the *Save* icon, and also click on the pencil to exit editing mode.

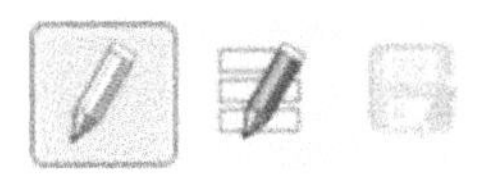

Now run the *Sigma* tool again for this area attribute:

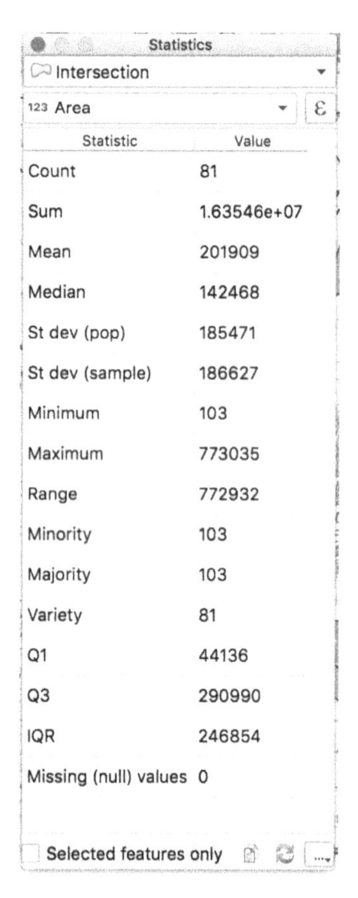

Take a look at the descriptive statistics for the sum of the total area, min, max, range, etc. Now do the same for the *$perimeter*, and you will see the perimeter measurements for each polygon.

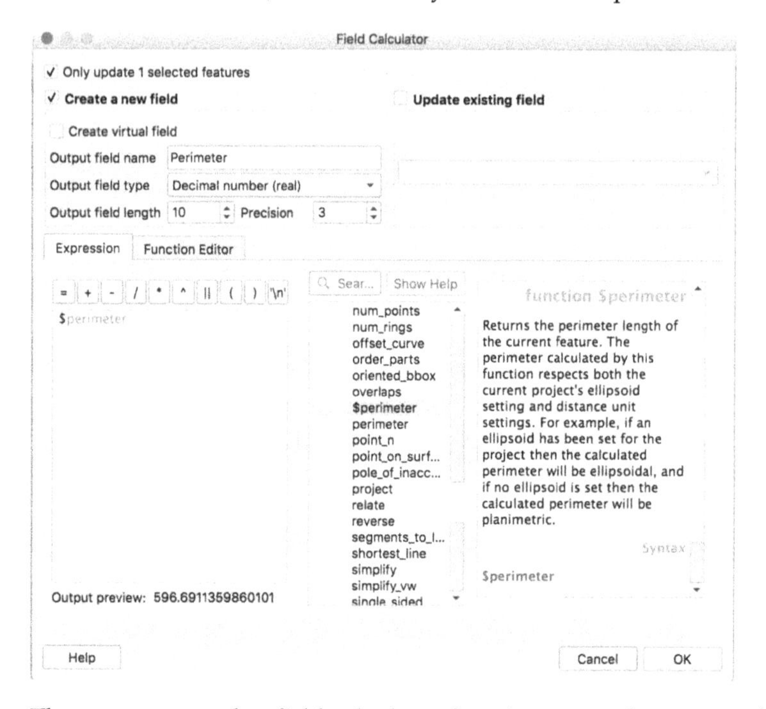

There are many other field calculator functions to explore, so read the documentation and learn more on your own.

6.7 Vector Boolean Operators

The other Boolean operators are variations on this same theme. Explore these to learn more. Go online and search, look for QGIS YouTube videos, or go to the QGIS documentation here: `https://docs.qgis.org/3.16/en/docs/training_manual/`

6.8 Points in Polygons

Another common vector function is *Point In Polygon*. This tells you how many individual points, such as mines, archaeological sites, or other point features, are in each individual polygon in a vector polygon dataset. Let's see how many of our `spothgt` points are within this intersection dataset. Choose `Vector->Analysis Tools->Count Points in Polygon` from the menu.

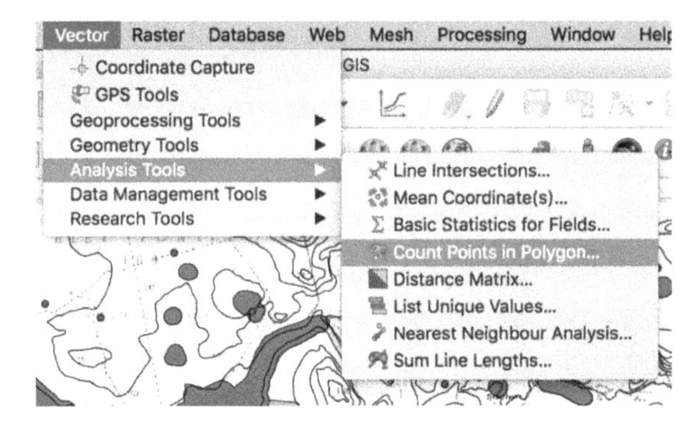

Enter the polygon and point layer names and *Run* the analysis.

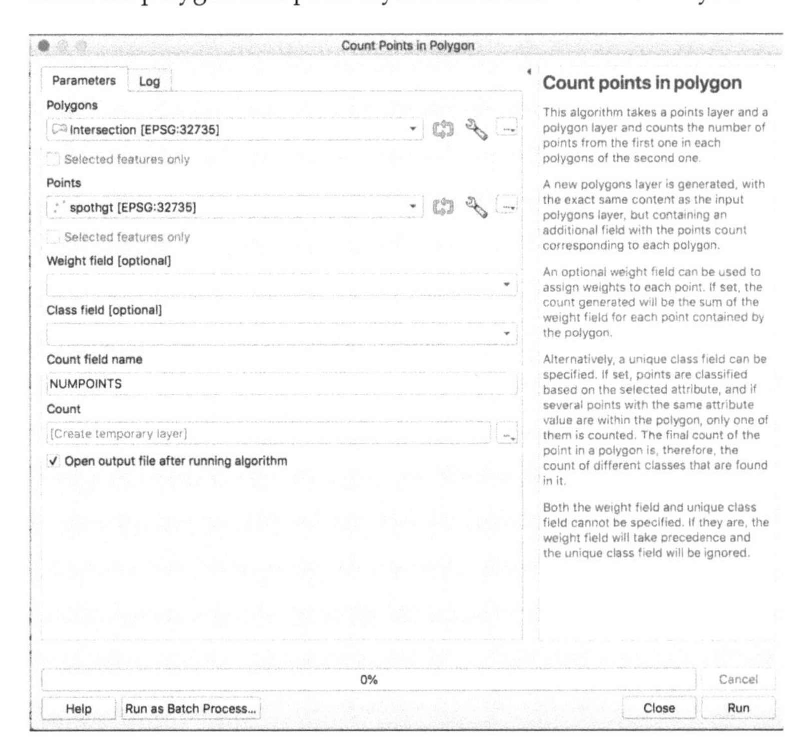

When complete, look at the results by clicking on the attribute table of the new scratch layer named `Count` and examining the *NUMPOINTS* field.

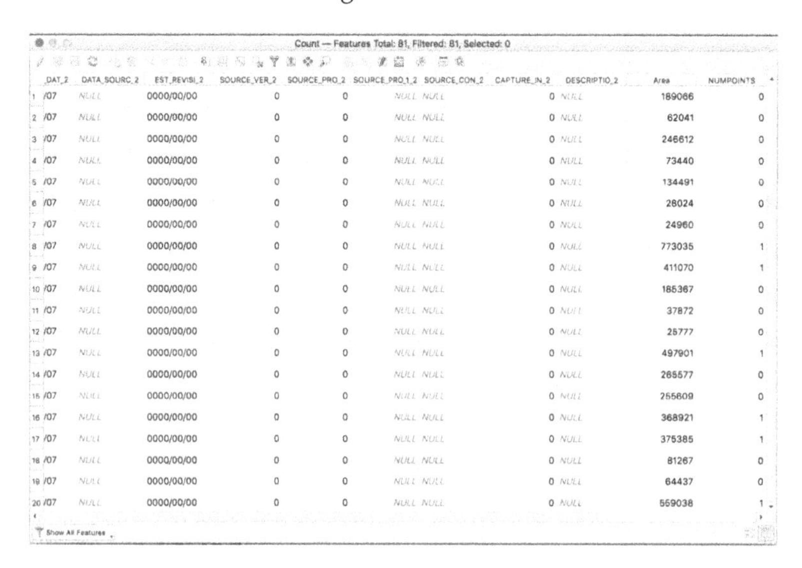

Count — Features Total: 81, Filtered: 81, Selected: 0

	DAT_2	DATA_SOURC_2	EST_REVISI_2	SOURCE_VER_2	SOURCE_PRO_2	SOURCE_PRO_1_2	SOURCE_CON_2	CAPTURE_IN_2	DESCRIPTIO_2	Area	NUMPOINTS
1	/07	NULL	0000/00/00	0	0	NULL	NULL	0	NULL	189066	0
2	/07	NULL	0000/00/00	0	0	NULL	NULL	0	NULL	62041	0
3	/07	NULL	0000/00/00	0	0	NULL	NULL	0	NULL	246612	0
4	/07	NULL	0000/00/00	0	0	NULL	NULL	0	NULL	73440	0
5	/07	NULL	0000/00/00	0	0	NULL	NULL	0	NULL	134491	0
6	/07	NULL	0000/00/00	0	0	NULL	NULL	0	NULL	28024	0
7	/07	NULL	0000/00/00	0	0	NULL	NULL	0	NULL	24960	0
8	/07	NULL	0000/00/00	0	0	NULL	NULL	0	NULL	773035	1
9	/07	NULL	0000/00/00	0	0	NULL	NULL	0	NULL	411070	1
10	/07	NULL	0000/00/00	0	0	NULL	NULL	0	NULL	185367	0
11	/07	NULL	0000/00/00	0	0	NULL	NULL	0	NULL	37872	0
12	/07	NULL	0000/00/00	0	0	NULL	NULL	0	NULL	25777	0
13	/07	NULL	0000/00/00	0	0	NULL	NULL	0	NULL	497901	1
14	/07	NULL	0000/00/00	0	0	NULL	NULL	0	NULL	265577	0
15	/07	NULL	0000/00/00	0	0	NULL	NULL	0	NULL	255609	0
16	/07	NULL	0000/00/00	0	0	NULL	NULL	0	NULL	368921	1
17	/07	NULL	0000/00/00	0	0	NULL	NULL	0	NULL	375385	1
18	/07	NULL	0000/00/00	0	0	NULL	NULL	0	NULL	81267	0
19	/07	NULL	0000/00/00	0	0	NULL	NULL	0	NULL	64437	0
20	/07	NULL	0000/00/00	0	0	NULL	NULL	0	NULL	559038	1

Show All Features

You can see that 72 of these points were within that intersection of buffers.

6.9 Union

Now we can use the *Vector->Geoprocessing Tools->Union* function to see all areas of the two buffers we created earlier. The dialog box looks like this:

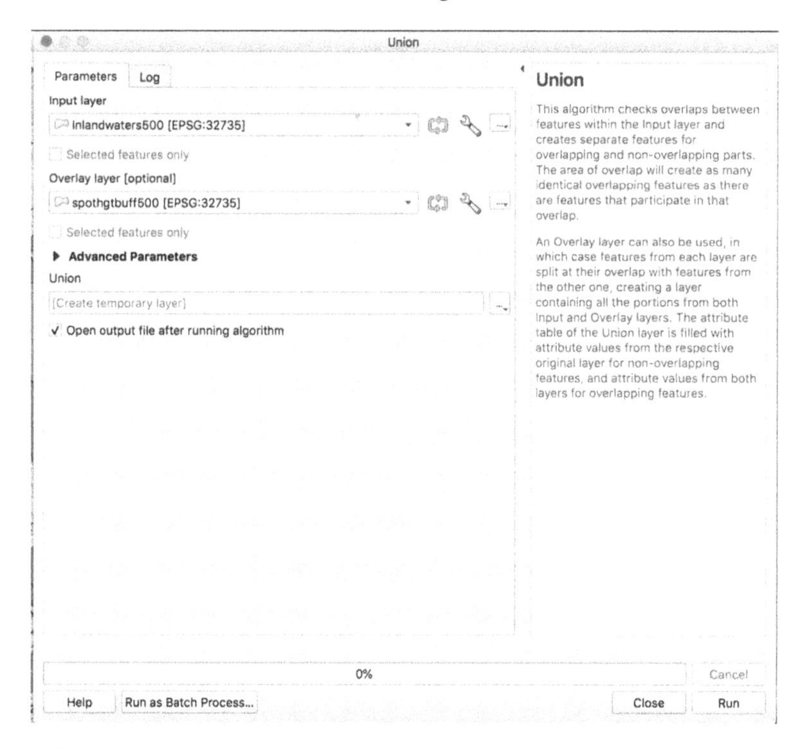

When we click *Run*, the result is:

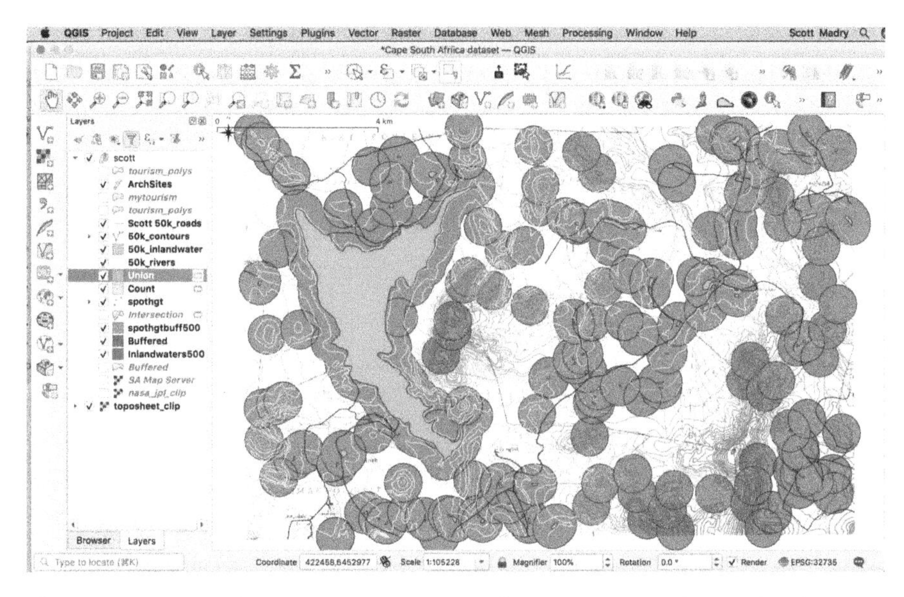

There are many other vector processing functions to explore. Try them out and learn. These are some of the most commonly used vector GIS processing functions, and QGIS, along with GRASS and other associated tools, have enough capabilities to do most of what you will ever want to do to process and analyze the relationships between various vector GIS data.

6.10 Vector Research Tools

Click on `Vector->Research Tools->Random Points in Polygon`.

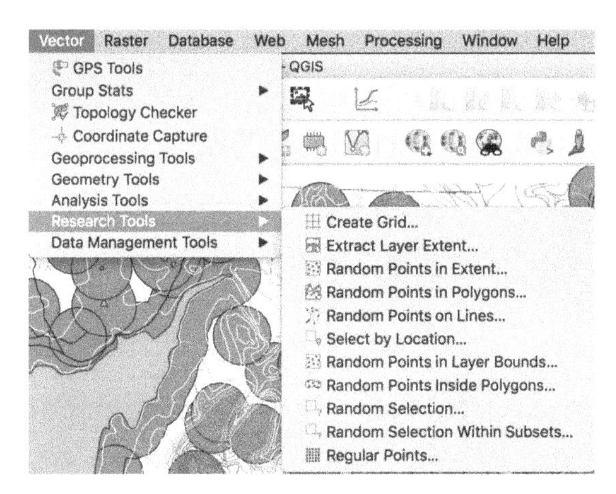

Fill in the options to have 100 points as shown below:

Pick the `50kinlandwaters` layer for the input layer, and it will create a new points file, and tell it to create 100 random points within the boundary layer. Click *Run* and it will run the process and add the new layer. You just created a scratch layer with 100 random locations on the landscape—very useful for random sampling, spatial modeling, etc.

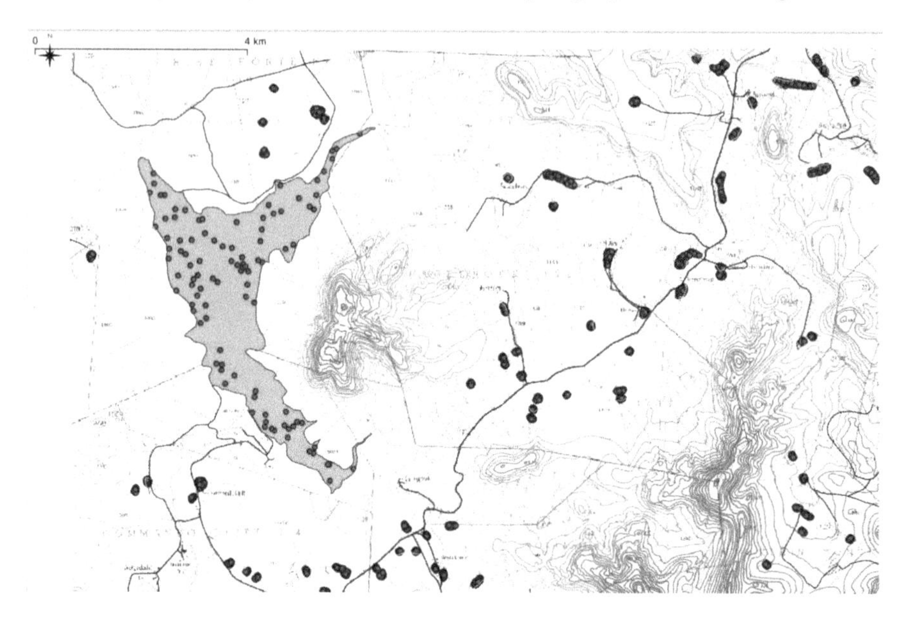

Right click on the new random points name in the *Layers* panel and click on *Properties*. You can change the symbol, size, add attributes, etc. Click on *Symbology*, and you can choose one of several icons and colors.

Note that you can also create a stratified sample of points based on the values of an existing vector layer. So you could have 100 points distributed by the percentage areas of a vector soils or a vegetation layer. We can do the same along a vector line such as roads (shown below) or streams:

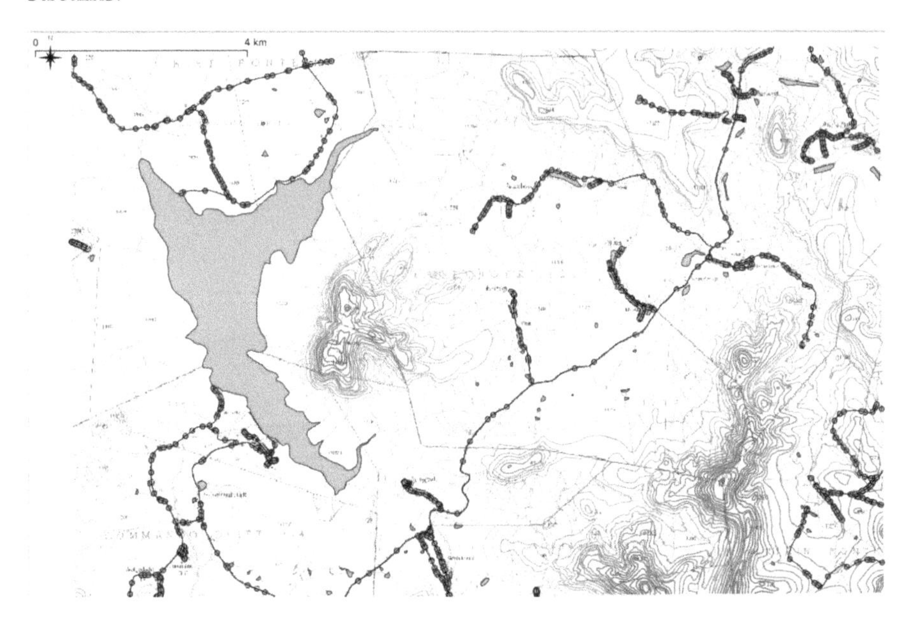

There are many other research options available to you:

6.11 Voronoi Polygons

Creating Voronoi polygons creates areas around point files, and are often used to create areas of influence. Go to the `VECTOR` menu and choose `Geometry Tools->Voronoi Polygons`.

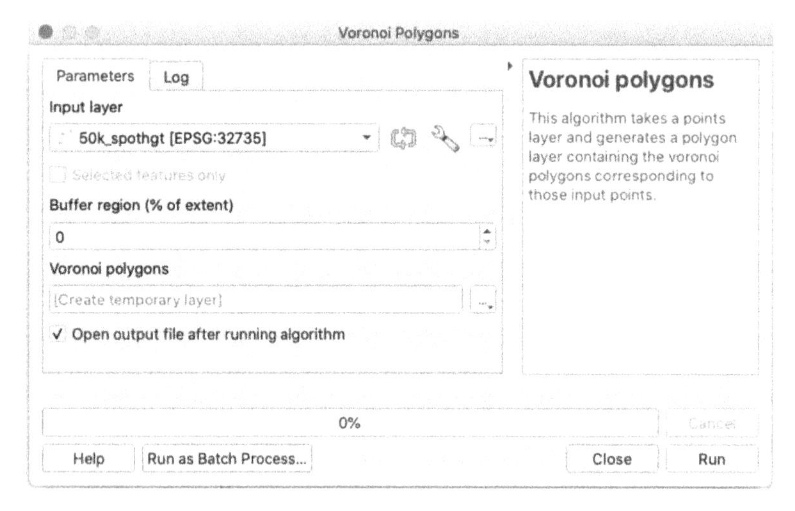

Use the point file `spothgt` (or a point file of your choice) as the input vector. It will only show you the point layers that you have open. Clicking *Run* will create a new scratch layer named `Voronoi polygons`. Put your `scott50kspothgt` layer above it in the *Layers* panel and you will see the result—equal area polygons around the *area of influence* of each point.

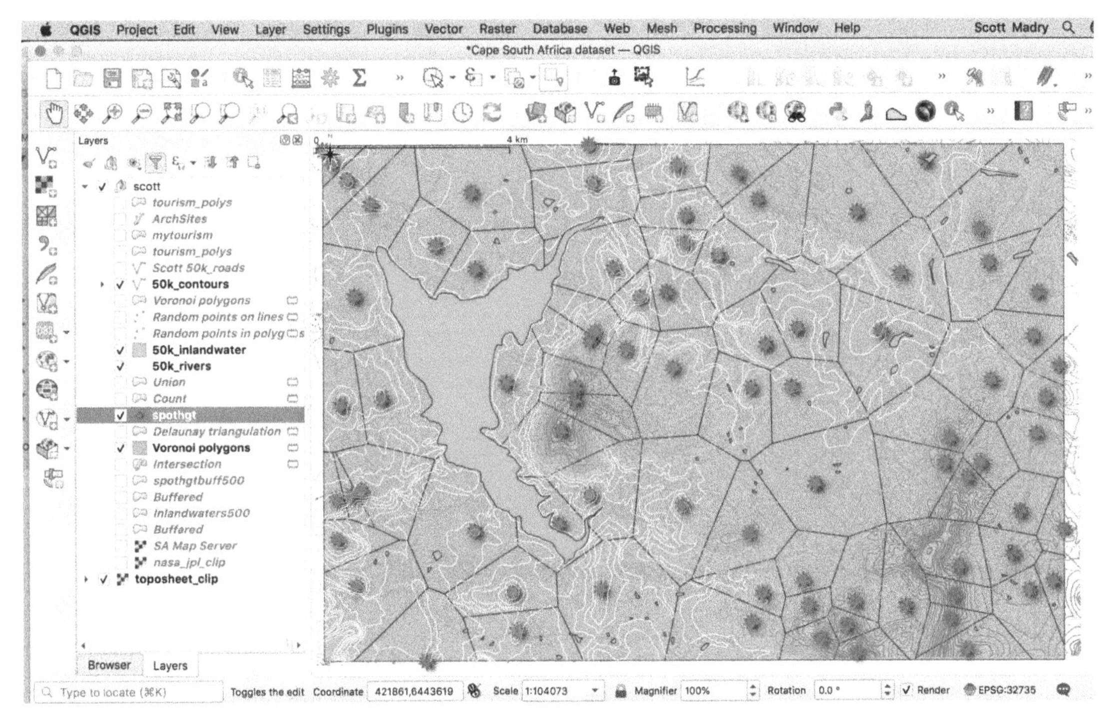

Remember to save your project often, just click `CTRL-s` in Windows or `command-s` for Macs.

6.12 Vector Delaunay Triangulations

Now do the same with the Delaunay Triangulations. Go to `Vector->Geometry Tools->Delaunay Triangulations`, and fill in the dialog:

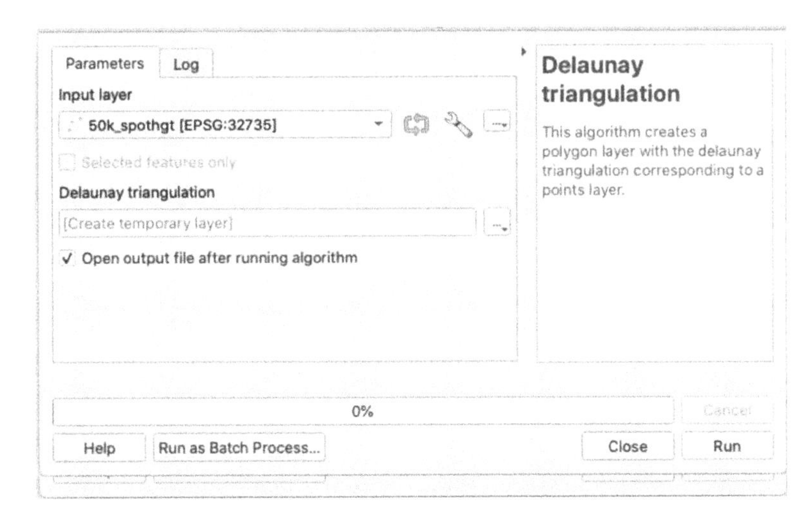

Pick the same file and click on *Run* at bottom right to create the triangulations:

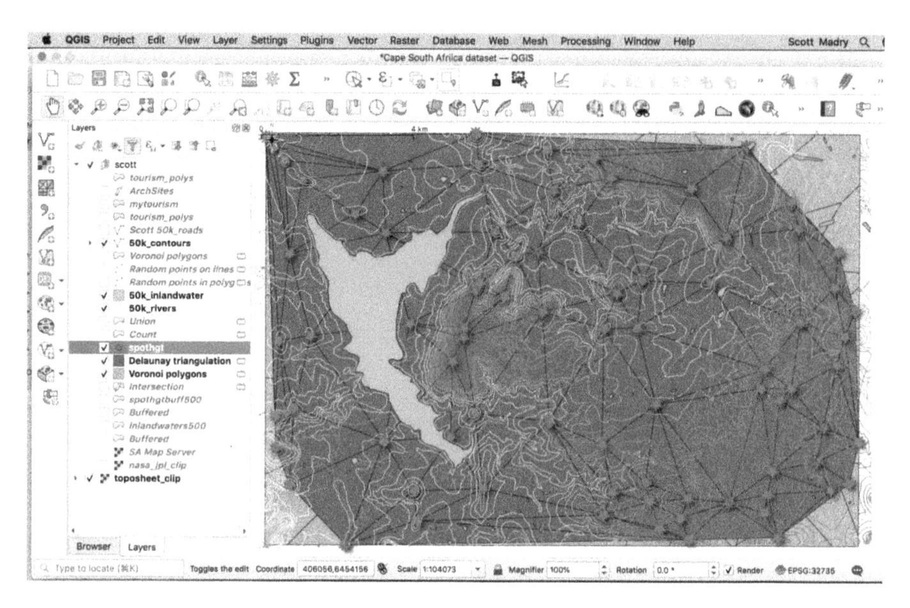

You can see this is the opposite of the Voronoi diagram, creating a TIN, or Triangulated Irregular Network, using the points as the nodes.

6.13 Recategorizing Vector Data

Now recategorize the color of each Voronoi polygon by clicking on the colored icon to the left of the layer name in the *Layers* panel, or right-click on the filename and go to *Properties->Symbology*. Set it to *Categorized* and for *Value*, use *POINTB*. Click on *Classify* and click *OK*.

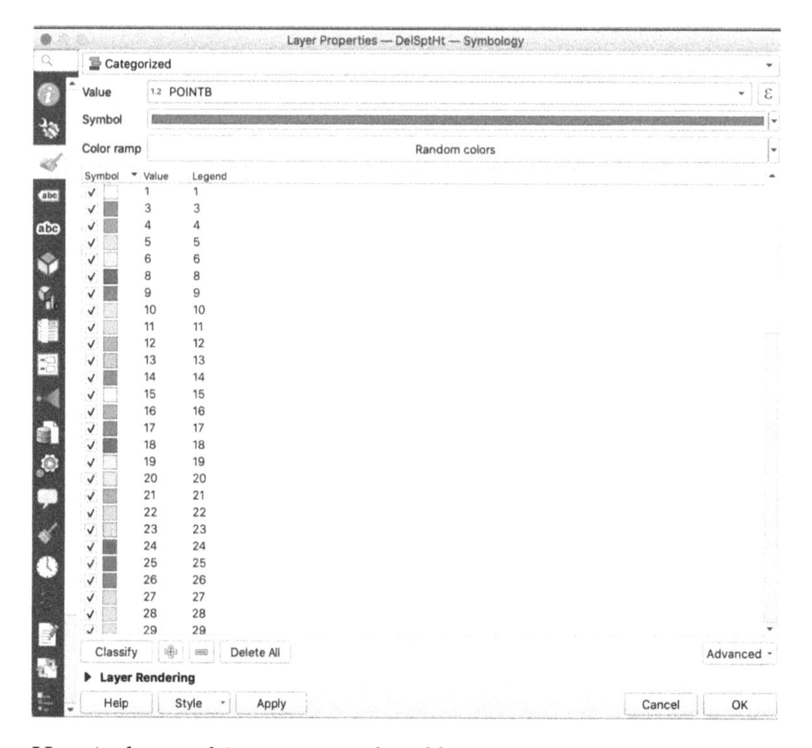

Here is the resulting map, rendered by category:

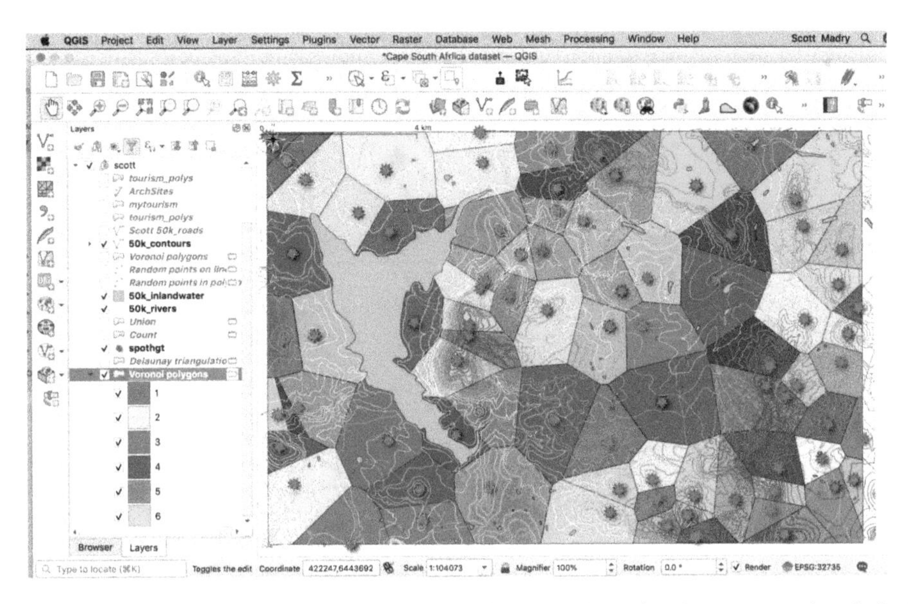

Note that each polygon's color is now displayed in the *Layers* panel at left. You can collapse this by clicking on the small triangle next to the check mark.

6.14 Vector Data Analysis

Under the `Vector->Analysis Tools` menu you will see several useful tools.

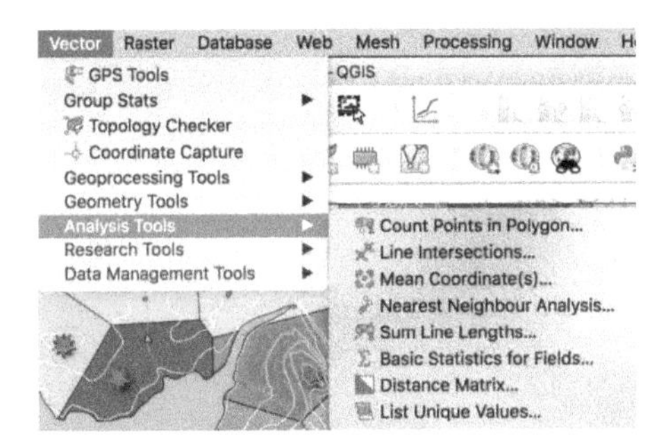

As we did previously, the `Count Points in Polygon` will tell you how many points in a file are in each category of a layer. The `Basic Statistics for Fields` is also useful. Your point layer `scott 50k_spothgt` contains hill and mountain peak elevations. Do the basic statistics for this, using *HEIGHT* as the target field:

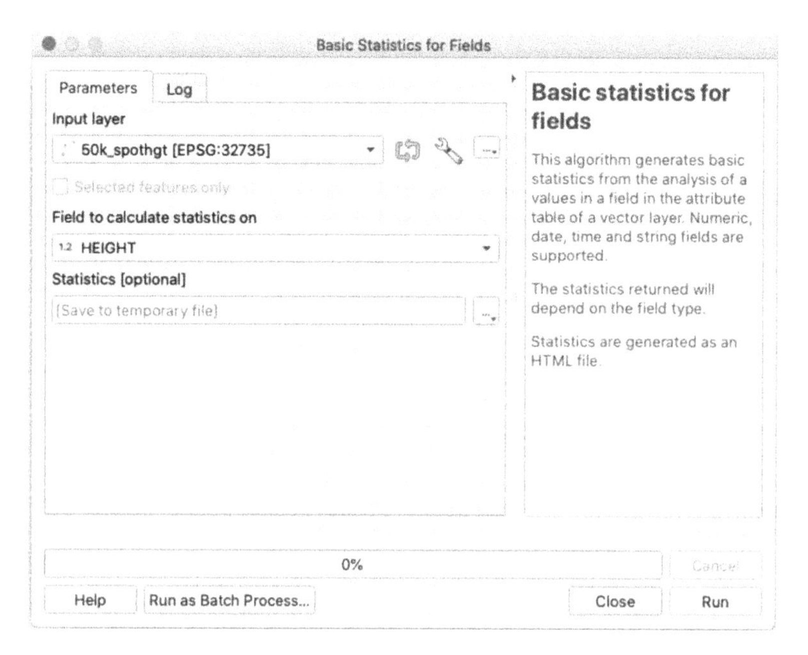

Again, if you click on *Log*, you will see the actual process that was run:

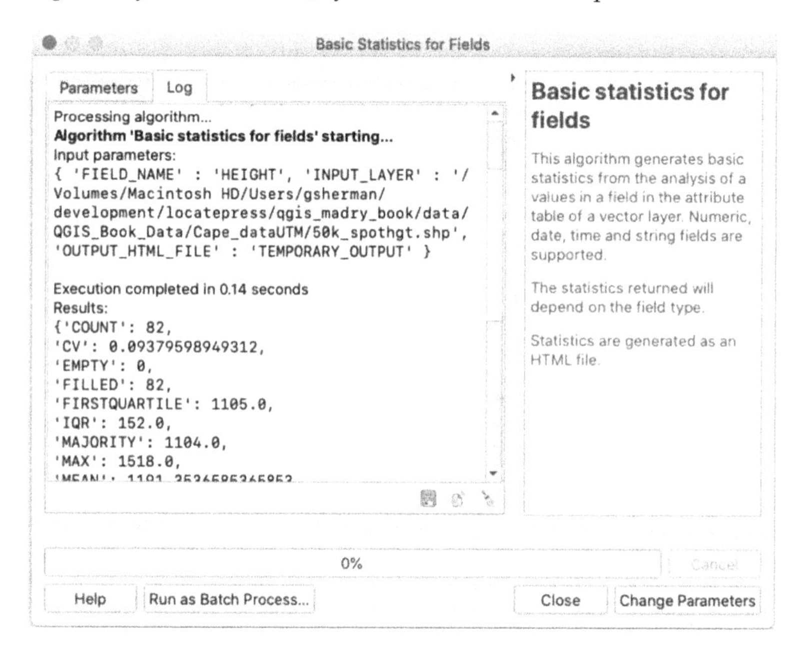

The results are stored in an HTML file and can be viewed by double-clicking on *Statistics* entry in the *Results Viewer* panel. Here are the basic statistics for the *HEIGHT* field:

```
Analyzed field: HEIGHT
Count: 82
Unique values: 75
NULL (missing) values: 0
Minimum value: 973.0
Maximum value: 1518.0
Range: 545.0
Sum: 97691.0
Mean value: 1191.3536585365853
Median value: 1186.5
```

```
Standard deviation: 111.74419523868762
Coefficient of Variation: 0.09379598949312
Minority (rarest occurring value): 973.0
Majority (most frequently occurring value): 1104.0
First quartile: 1105.0
Third quartile: 1257.0
Interquartile Range (IQR): 152.0
```

There are many more capabilities to be found in the Vector menu, so feel free to explore and discover the many tools available.

7. QGIS Plugins

QGIS is part of the OSGeo stack of open source geospatial tools, and so there are many additional capabilities available that are actually outside the core QGIS software. One of the reasons that QGIS has become so popular is its ability to accommodate plugins, allowing others to share their work and developments with the larger QGIS and open source communities.

There are many QGIS plugins available now, over 1,300, written and submitted by developers around the world (see `https://plugins.qgis.org/plugins/` for the complete current list). The website lets you see all the plugins, new ones, most popular, most rated, etc. You can also search for plugins by topic. Several plugins are automatically loaded when you install QGIS.

7.1 Core Plugins

These are referred to as *core* QGIS plugins, and they are really a part of the core capabilities of the system. QGIS continually grows in this way, adding new core functions by incorporating stable and useful plugins that expand the functionality of the software.

Many other plugins are available for you to download—all free. These provide many additional features that are available to you, and you can pick and choose which ones you want, load one and use it and remove it, etc. They range from the small to very large, such as the Semi Automated Classification satellite remote sensing plugin, which is a fully capable image processing system for QGIS.

Another very powerful plugin is the InaSAFE disaster planning and modeling tool (inasafe.org). This adds a significant disaster planning and management capability to QGIS. Many plugins are discipline-specific, such as for archaeology or ecology, and others are national, working with specific data from France or Portugal, for example.

7.2 Accessing the QGIS Plugin Repository

The website `https://plugins.qgis.org` is the QGIS plugin web portal. To download plugins, go to QGIS `Plugins` menu and choose `Manage and Install Plugins`.

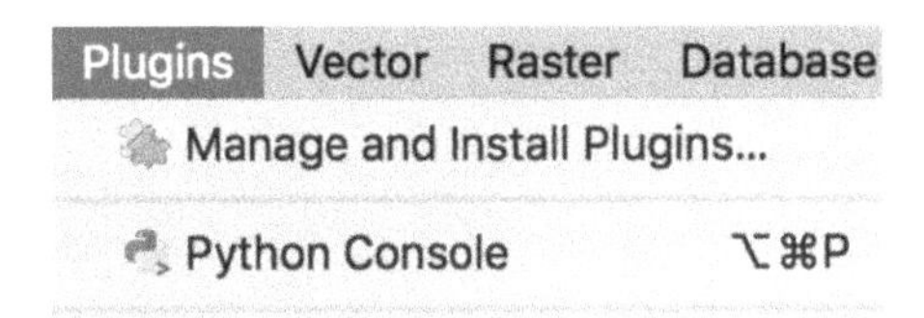

7.3 The QGIS Plugin Manager

The plugin manager is where you can find and install plugins into QGIS. Some will be automatically loaded for you, as they are considered to be core capabilities, but there are many more. Open the *Plugin Manager* and you will see:

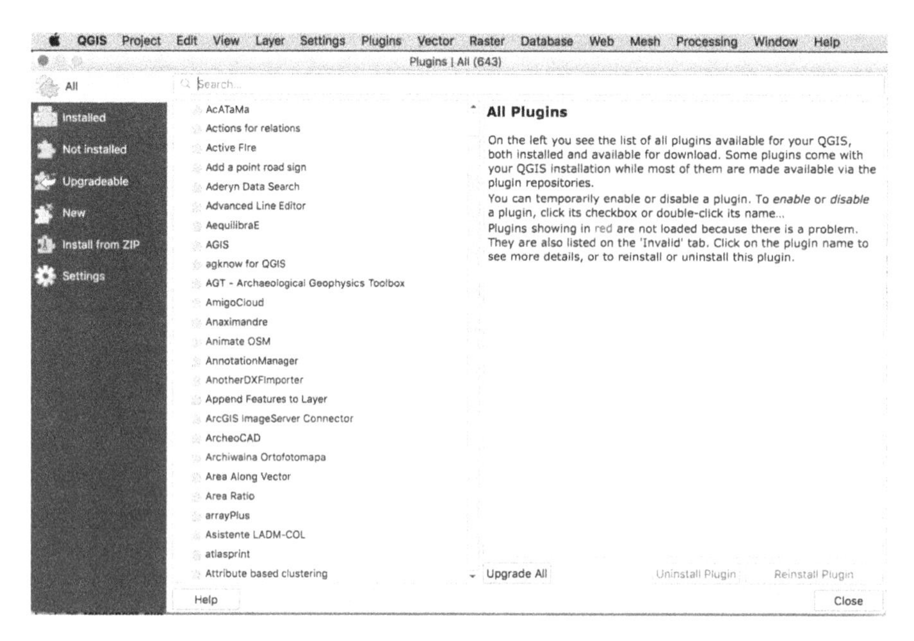

On the left are tabs, *Installed*, *Upgradeable*, *New*, etc. Click on *Installed* to see your installed plugins, including the core plugins. You can check or uncheck them to activate them. Some QGIS plugins come already installed, but you have to check them to make them active. Once you click to download a new one, a new menu, toolbar, or icon is added to the QGIS interface and becomes available to use. You must be connected to the Internet to access the remote repositories and download plugins.

To update your current plugins, click the *Upgrade All* button at the bottom. To locate and load additional ones, click on the *All* tab to see the list.

Click on *Installed*, and you will see all of the plugins that are currently installed. Many plugins will automatically load when you install QGIS (I will have more than you, as I have downloaded several):

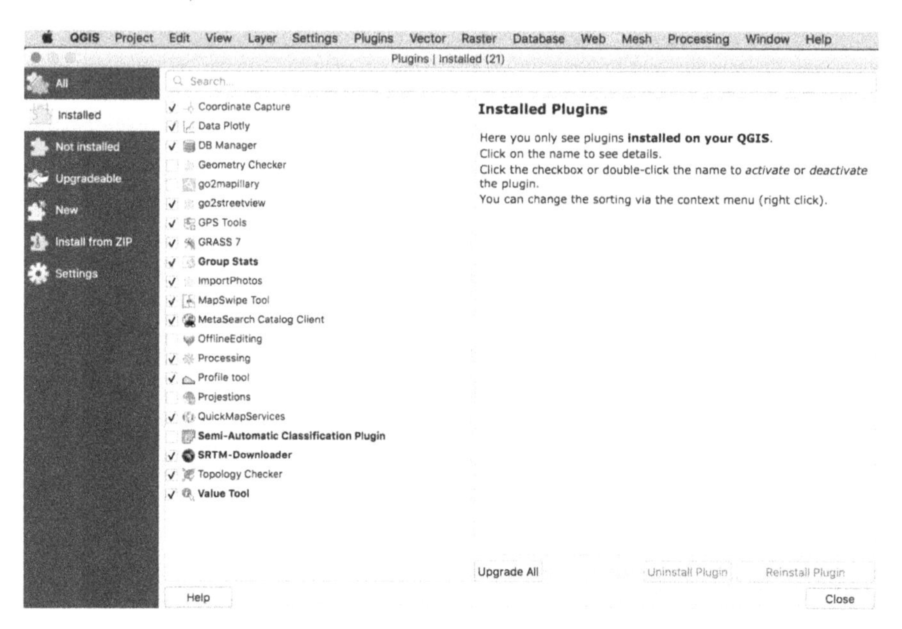

It is important to note that you must activate many of these, by checking the box. If you can't find a plugin in the user interface, go here and check if it is enabled.

7.4 Plugin Settings

If You Click On *Settings* at left, you will see the plugin repositories available and can adjust other settings.

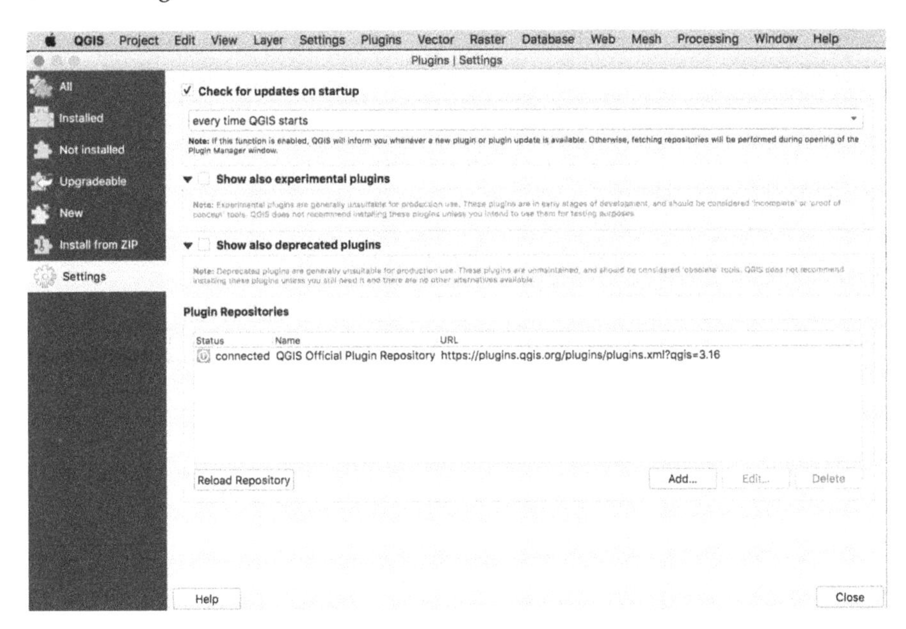

It is a good idea to check the top box, *Check for updates on startup*, to check for new updates each time you start QGIS. Be aware that plugins in the official repository have been tested and are stable, while other individual repositories contain plugins of varying quality and maturity. Be wary of experimental plugins. I would suggest working with the official repository first.

7.5 Searching for Plugins

There is a search bar at the top of the *Plugin Manager*. If you click on a plugin, you can see a description, a rating (stars) and downloads. Here I searched for OSM (OpenStreetMap), and got a list of several plugins:

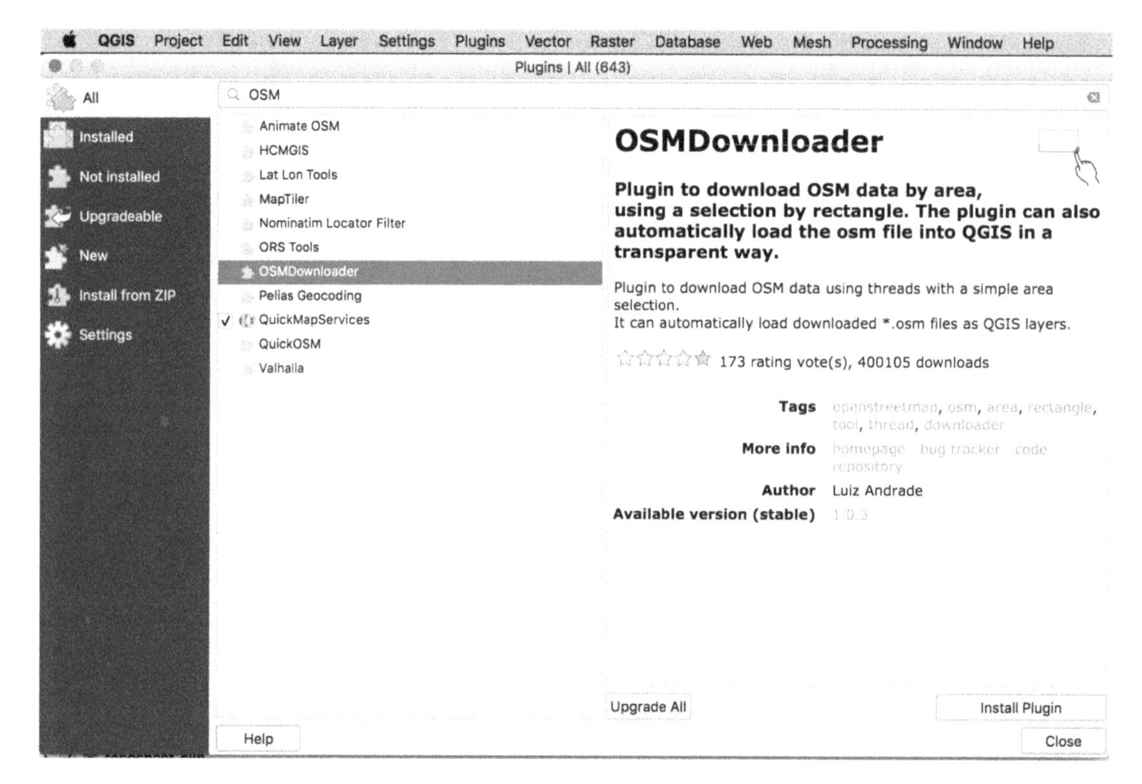

You can see the *OSMDownloader* plugin has a high rating. You can click on version number to open the information page in your web browser. Search for CAD, or topology, or Belgium, etc. or simply click on ALL at left and scroll down. There are more every month, so check back often.

Now let's look at a few of the really useful plugins.

7.6 QuickMapServices

This is one great plugin! It allows you to *live link* many different datasets from around the world in your QGIS display. These include satellite imagery and background maps from Google, Apple, ESRI, NASA, and many more.

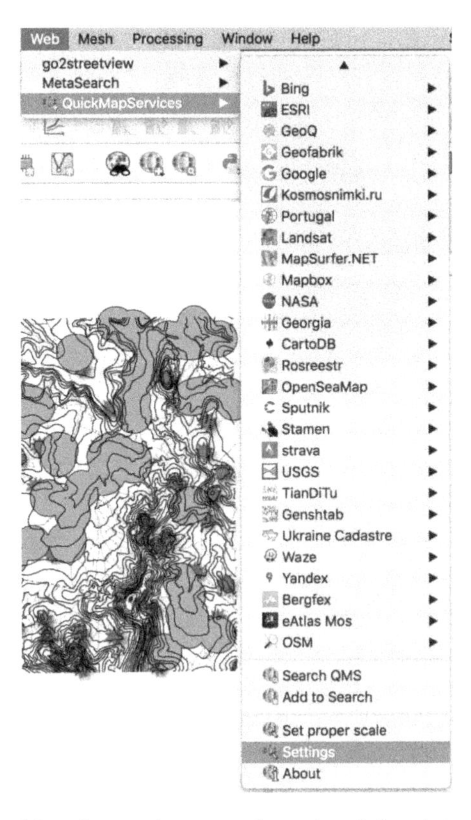

You do not have to download the data, and the chosen dataset will automatically overlay correctly, zoom, and pan as you explore your data. Once you download the plugin, it will appear under the `Web` menu along the top of your QGIS window. To add services, choose to `Web->Quickmapservices->Settings` from the menu, click on the *More Services* tab and click on *Get contributed pack* at the bottom.

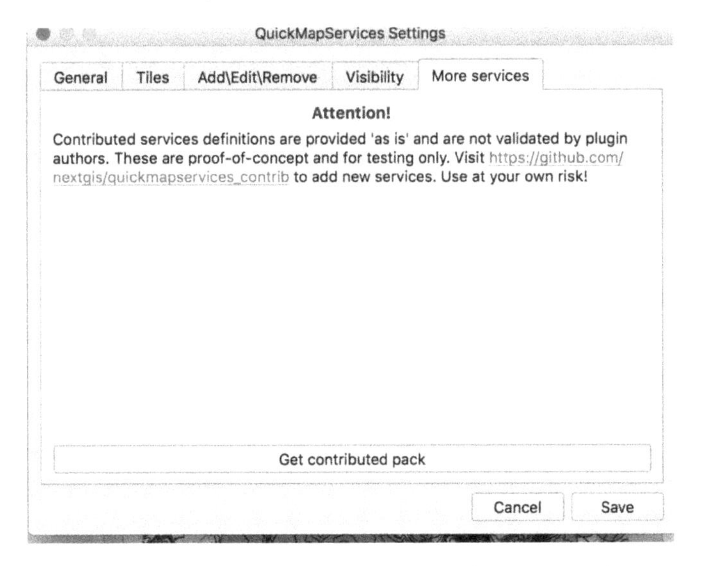

You can disable services you are not likely to use by going again to `SETTINGS` and clicking on the *Visibility* tab.

Click on *Save* and you are ready. Now click on `Web->Quickmapservices->Bing->Bing Satellite`

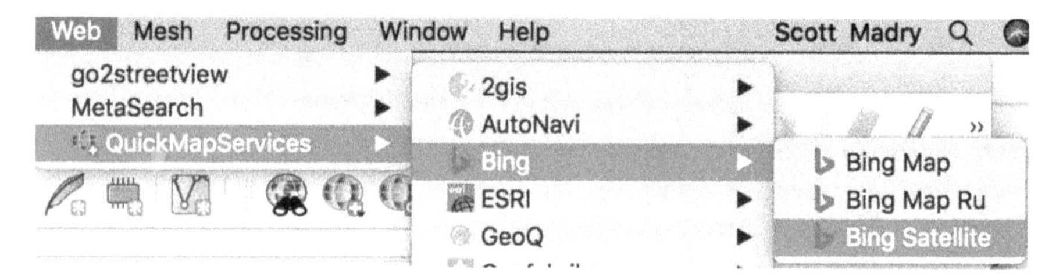

A satellite background image will be added to your map, as shown below:

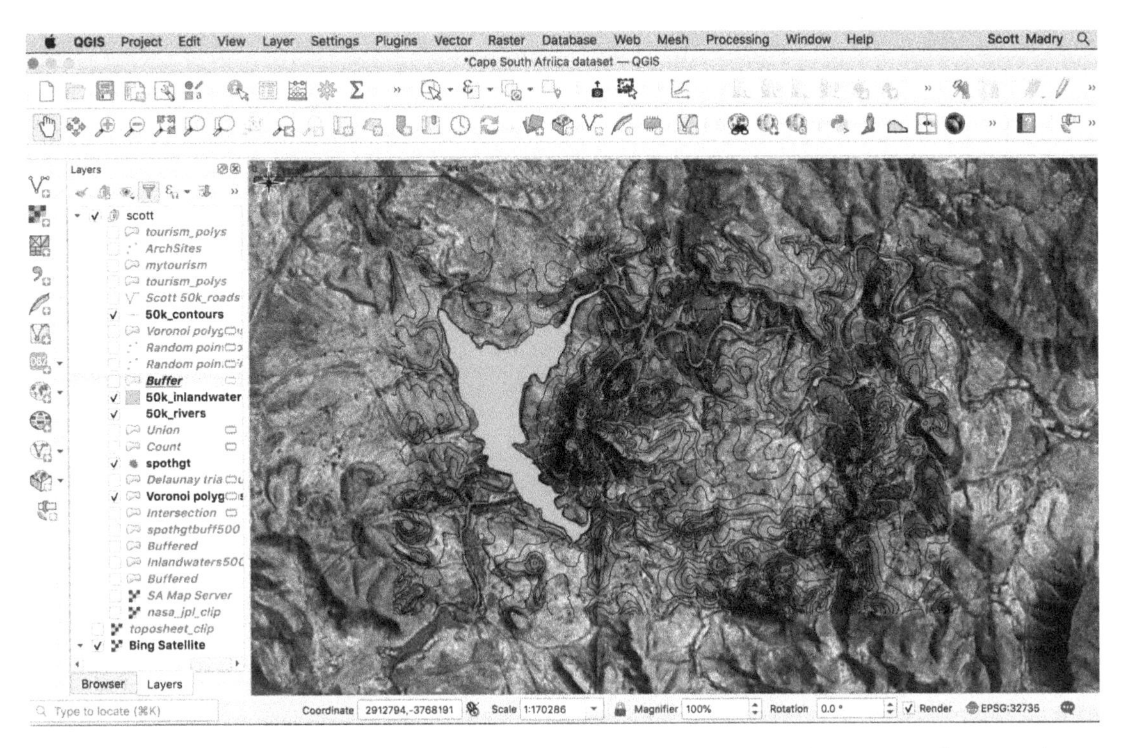

The Bing Satellite image is added at the bottom of your *Layers* panel. Zoom and pan and you will see that the image follows you.

Try several other datasets. Here is the ESRI shaded relief data for our study area:

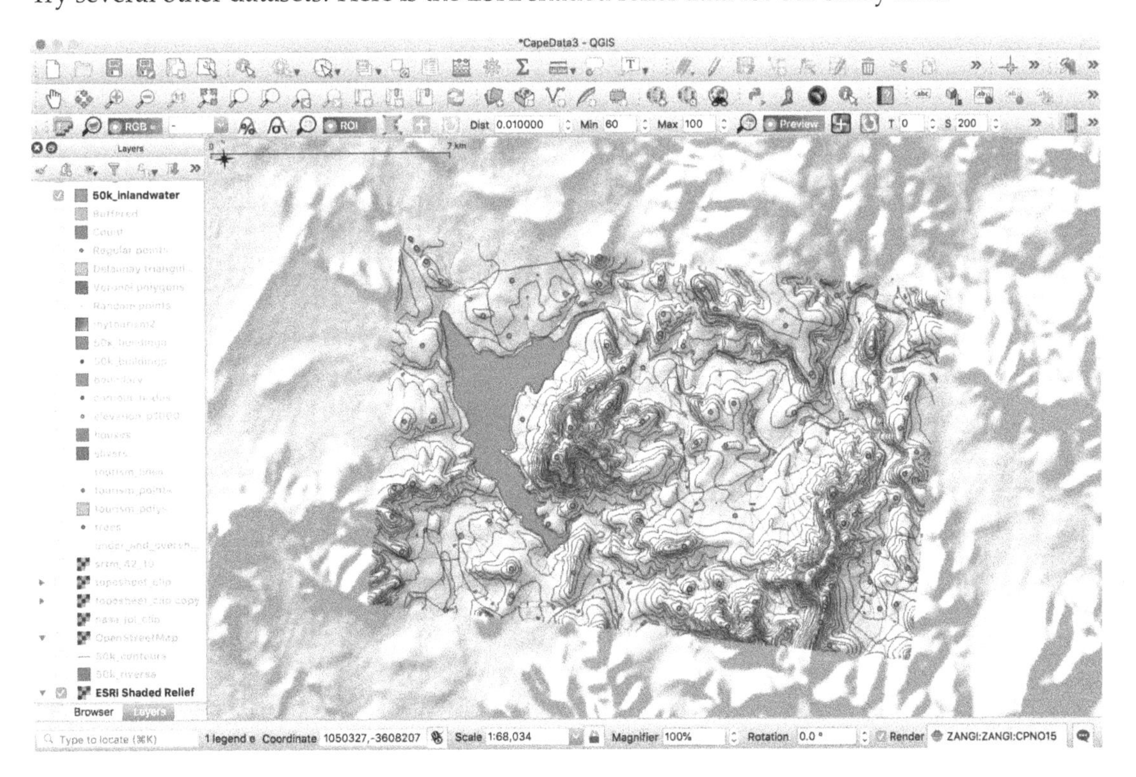

7.7 AutoSaver

QGIS does not automatically save your data, and this can be a problem. The AutoSaver plugin that will do this for you. It has over 90,000 downloads and a 96% positive rating, so you should probably download it.

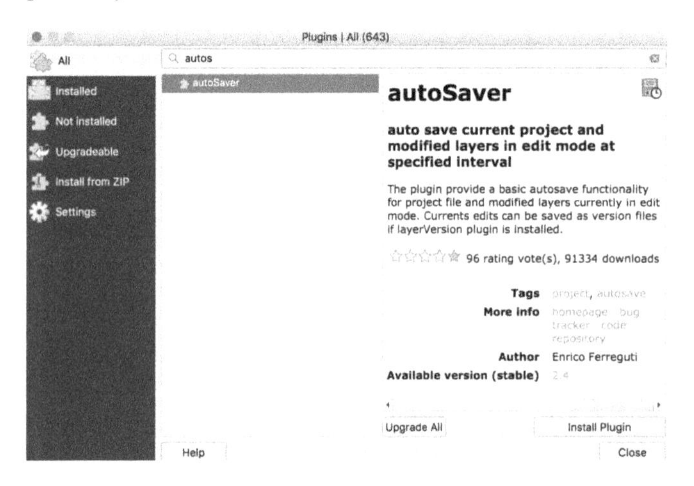

7.8 Coordinate Capture

This plugin allows you to point and click on the map and see or save the coordinate pair in both your current project CRS and another. First, ensure that the plugin is enabled, by going to `Plugins->Manage and Install Plugins` and typing "coordinate" at the top. Click the box to the left to enable it. This adds the plugin to both the Vector menu and toolbar. Click on `Vector->Coordinate Capture`. or click on the icon on the toolbar.

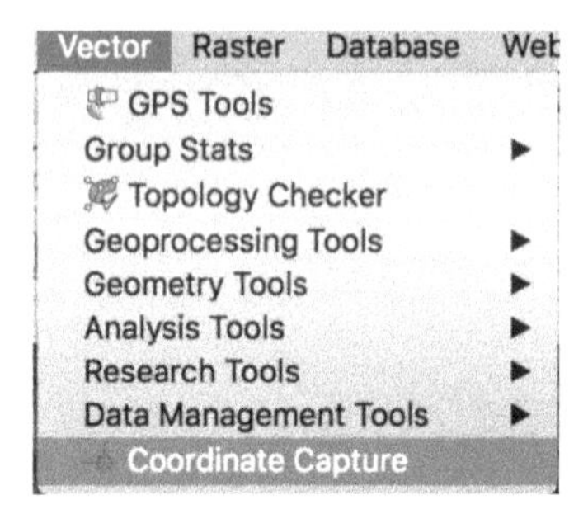

When you click on `Coordinate Capture`, a new panel is added below the *Layers* panel. Click *Start Capture* to begin collecting points, and click the copy icon (to the right of the coordinates) to copy individual coordinate pairs to the clipboard.

Click on the top button to choose a different CRS. Try WGS84, click on *Start Capture*, then click at different locations on the map. You will see both sets of coordinate pairs. This is useful for georeferencing imagery, locating ground points, and other operations. You can direct your output to a clipboard and save it.

7.9 GPS tools

This is a very powerful GPS toolkit. Again, go to the *Plugin Manager*, type *GPS*, then click the checkbox to activate it. This plugin is always loaded with QGIS, but you may have to activate it.

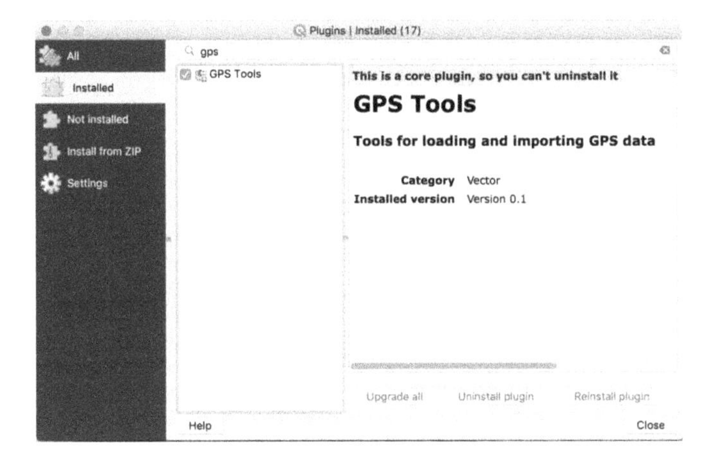

The `GPS Tools` now appear in the Vector menu, or you can click on the GPS icon to bring up the *GPS Tools* dialog. This powerful suite of tools lets you download GPS files into your GIS as point, line, or polygon vector files, upload point files to your GPS (very useful when wanting to navigate to an interesting location in the field—for predictive modeling, for example), and to convert data between the various GPS data formats.

If you use a Garmin or Trimble, you should be all set, as QGIS will directly recognize these files. Users of other GPS units have some massaging to do, but it works. QGIS uses the GPX (GPS eXchange) format, which is a standard interchange format for GPS data. GPSBabel is a free program (`http://www.gpsbabel.org`) that is a part of the QGIS download that allows data in other GPS formats to be converted into GPX. When you click on the `Vector->GPS Tools` menu you should see this:

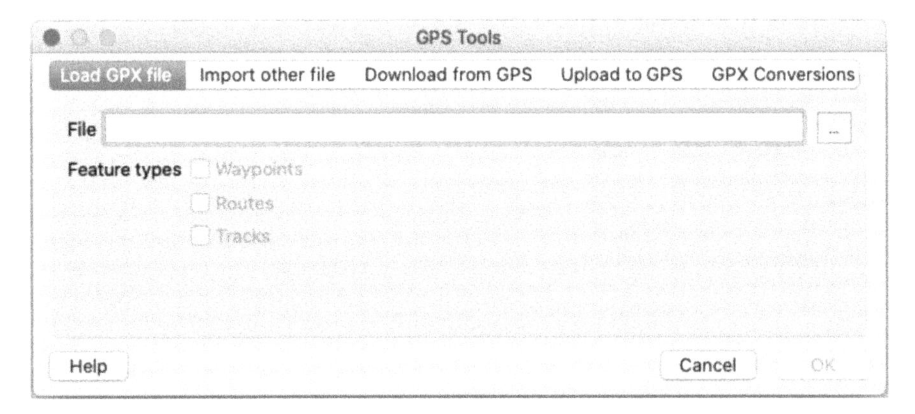

Using the five tabs across the top, you can load a GPX file, import GPS files, download files from a GPS, upload to your GPS, and do GPX data format conversions. It is very capable and robust, and a great example of how the GPSBabel tools were integrated into QGIS, rather than developing a new system when GPSBabel was already available.

7.10 Live GPS Tracking

There is also a capability in QGIS to show live, real-time GPS locations in your current GIS project. This is very useful when you are in the field with a mobile computer or using a laptop in a vehicle. To activate this, go to `View->Panels->GPS Information`. This will open the GPS panel shown below:

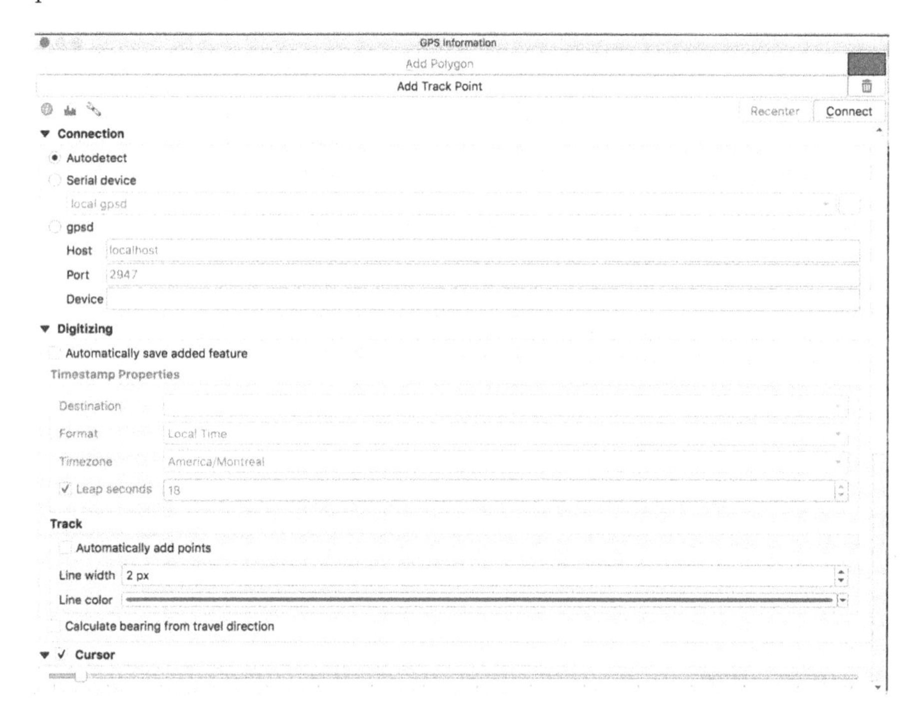

Note the three icons at top left. The right icon allows you to set up the connection to your live GPS unit. The center icon will display your signal strength and the location of current satellites in your attached GPS receiver as shown below:

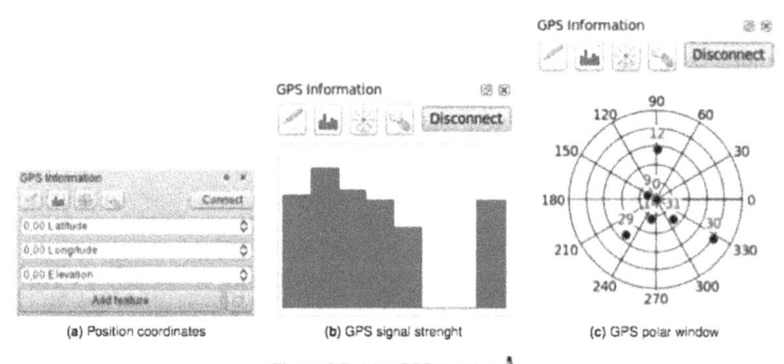

(a) Position coordinates (b) GPS signal strenght (c) GPS polar window

Figure 3.7.: Live GPS tracking

The icon on the left shows your GPS data (this screen is blank):

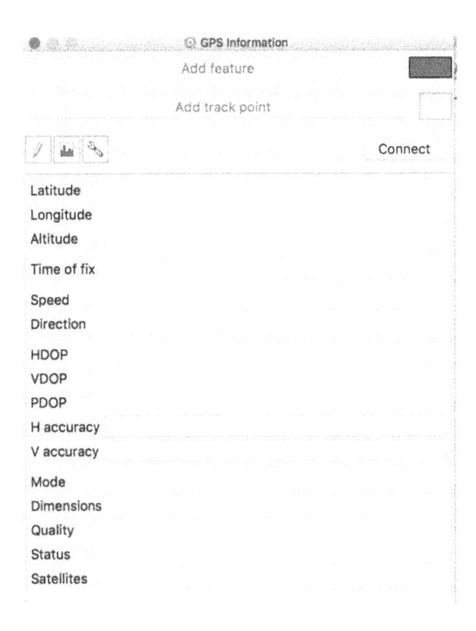

If you want to record your real-time position to the map canvas as a new GeoPackage or shapefile, you have to create a new vector layer first (`Layer->Create Layer->New Shapefile Layer`), make it a point, line, or polygon, save it, and then toggle editing (turn on editing). This is in order to be able to record your track. This lets us do real-time, one-to-one digitizing in the field of various point, line, and polygon features. This is a very powerful field tool.

7.11 InaSAFE- The Disaster Planning and Response Tool

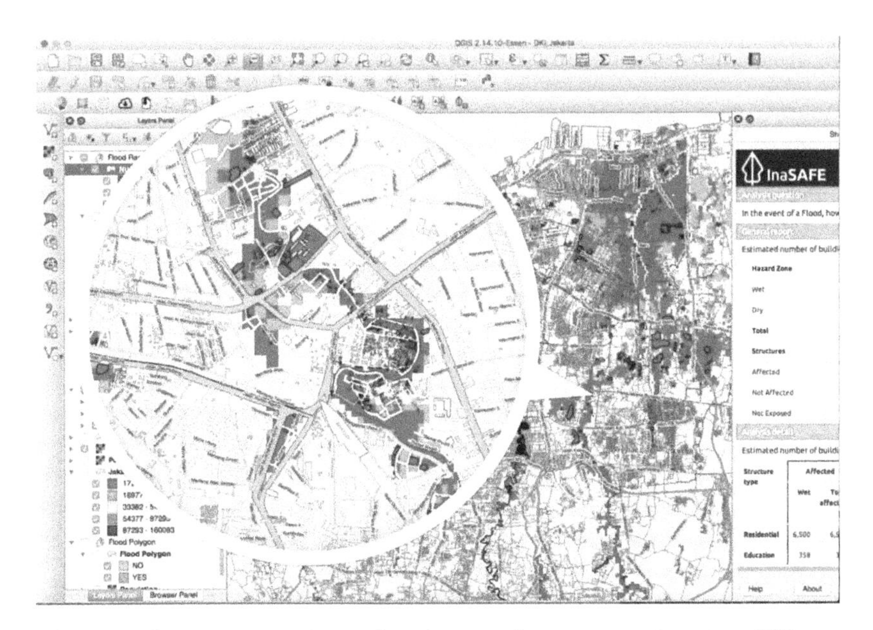

This is a disaster mapping, planning, and response planning GIS created to work inside the QGIS environment (`http://inasafe.org`). It was developed by the World Bank's Global Facility for Disaster Reduction and Recovery, the Australian Aid organization, and the Indonesian National Agency for Disaster Management. It lets you run simple and structured GIS analysis of floods, earthquakes, tsunamis, and volcanic eruptions, and model the number of people affected and their disaster needs. An InaSAFE screen is shown below:

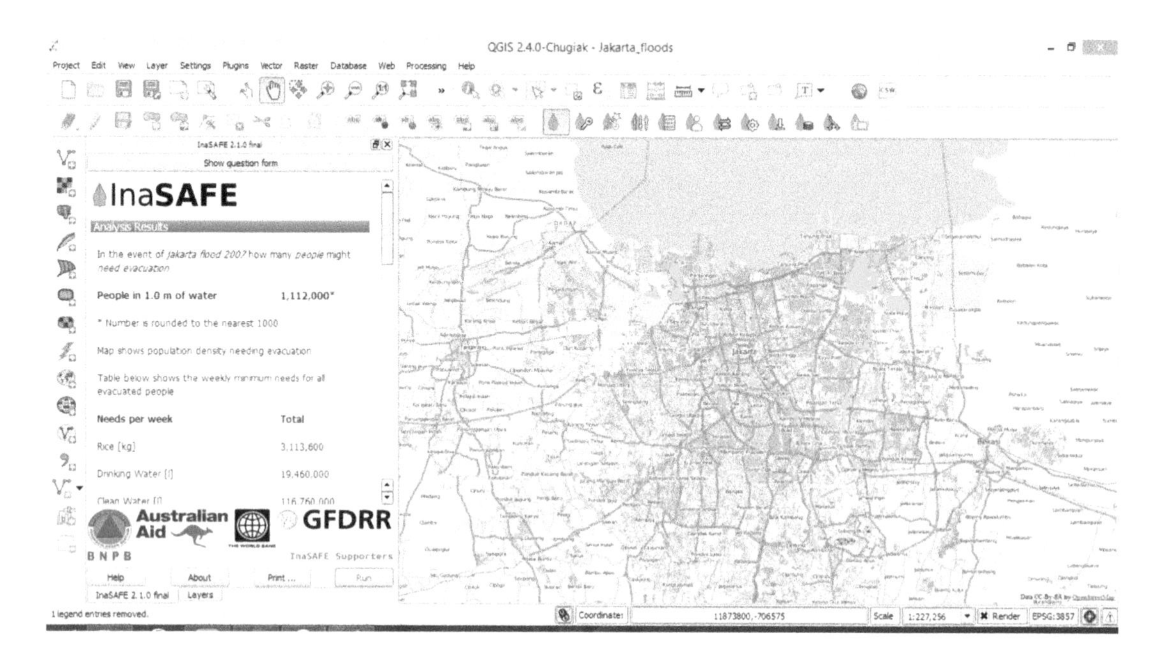

7.12 Semi Automatic Classification Satellite Remote Sensing Plugin

This is a very powerful but simple to use plugin that allows for the processing of satellite remote sensing images into useful QGIS data layers (`http://fromgistors.blogspot.com/p/semi-automatic-classification-plugin.html`). This takes you all the way from downloading free Landsat and European Sentinel and other satellite data, through the data processing, and generating final land use/land cover maps and statistics, as shown below:

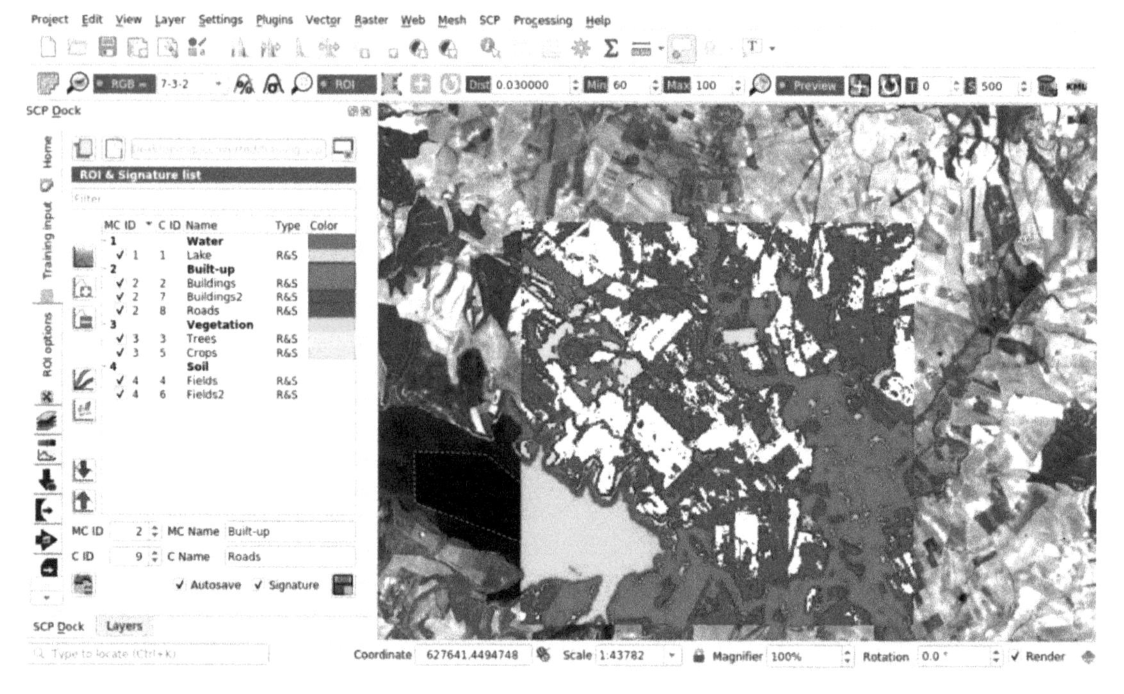

You can use the *Plugin Manager* to install and enable the plugin. Additional tutorials and data are available at the SCP website. The plugin adds a comprehensive satellite remote sensing capability to QGIS, and is well worth learning. This lets you download for free and work with

ASTER, GOES, Landsat, MODIS, Sentinel-1, Sentinel-2, Sentinel-3, imagery. This is the work of Luca Congedo of La Sapienza University in Rome, Italy.

7.13 Offline Editing

This very useful tool allows multiple people to work with a laptop or cell phone in the field and then sync with a master GIS database, allowing multiple users to keep their data in sync. Install and enable it from the *Plugin Manager*. It is very simple to use—add a vector layer to a QGIS project and then from the main menu, choose `Database > Offline Editing > Convert to Offline Project`.

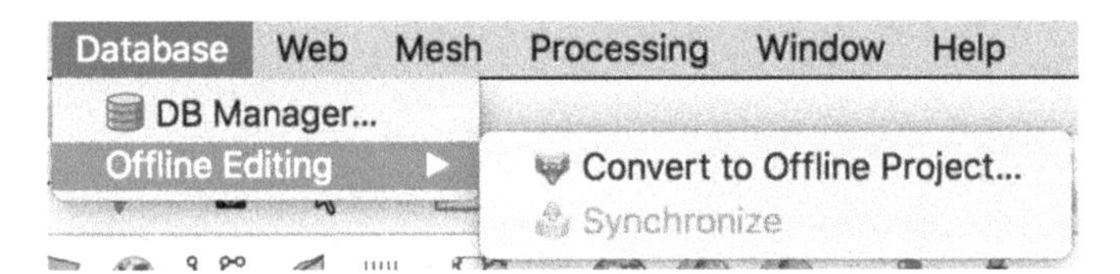

and choose which files are be saved. Then you can edit the data offline and to upload, you go `Database->Offline Editing->Synchronize`, and it will update that layer.

7.14 More Plugins

There are many, many more very useful plugins available to you. Plugins for drones and drone video, CAD, etc. Plugins provide additional capabilities and support to niche applications, such as archaeology, ecology, urban planning, and wildlife management, or for nonstandard national data formats or datasets. We can download the plugins we want, but the entire community is not required to install them, thus providing flexibility without making the overall system unwieldy.

There are so many other plugins, and more are added with each version of QGIS, so be sure to keep up to date and check the plugin repositories frequently. There is even a plugin (Plugin Builder) that creates a QGIS plugin template for you, so you can start developing your own. Go and look around and explore, and visit the repository regularly to both update your installed plugins, and find new and useful ones. Many plugins had to be rewritten in the conversion from QGIS 2 to 3, so check the repository often, as many are in the process of being updated and new ones added frequently.

If you are interested in contributing a plugin, and if you have something useful I hope you will, please see the QGIS Plugin Portal at: `https://plugins.qgis.org` to see the resources for plugin authors and information on how to create and publish your new QGIS plugin.

8. Raster Data and Analysis

Vector and raster GIS data are very different. Vectors are composed of points, lines, and polygons, while rasters are grid squares, like a chess board, with each square having a data value such as an elevation or vegetation value. Rasters and vectors are stored and analyzed differently in our GIS, and we can easily convert one into the other. But it is not a question of either/or, but rather how we use all of our data appropriately.

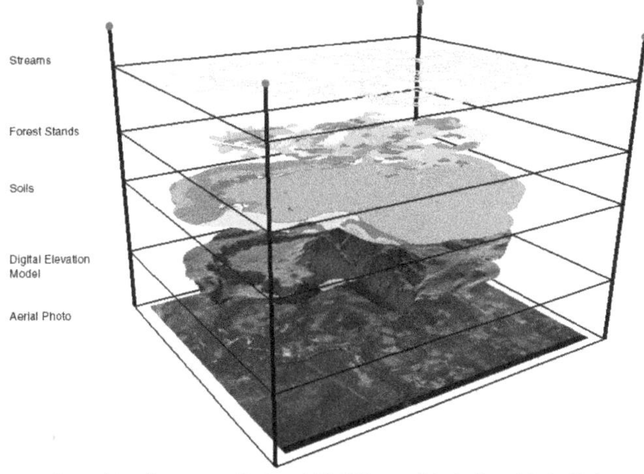

Source: https://courses.washington.edu/gis250/lessons/introduction_gis/index.html

8.1 Why Use Raster Data?

Raster GIS data and analysis are often preferred when working with terrain and other natural or environmental data such as elevation, slope, aspect, soils, land cover, hydrology, and drainage. Conversely, vector data is preferred for discreet boundary type data such as property ownership and urban data. Raster data are also used for satellite remote sensing and other kinds of aerial imagery, as well as for historical maps, which can be an excellent source of information about land changes over time.

Raster data are becoming much more commonly used in GIS and is often now freely available, and we can easily convert between the two formats. QGIS has multiple raster processing and analysis functions, which are accessed from the `Raster` menu and are mainly GDAL routines that have been incorporated into QGIS. The `Processing` menu provides access to routines that come from GRASS, SAGA, Orfeo, and other sources. The standard raster data format for QGIS is the TIFF or GeoTIFF (Georeferenced Tagged Image File Format). For additional information, see: `https://earthdata.nasa.gov/esdis/eso/standards-and-references/geotiff`

8.2 Using the GDAL Library in QGIS

The QGIS GDAL tools offer a GUI for the Geospatial Data Abstraction Library (GDAL), that are raster data tools that allow you to query, reproject, warp, and work with a wide variety of raster data such as remote sensing and raster DEM (Digital Elevation Model) files. You can create vector contour lines from raster DEMs, warp and georeference remote sensing images using ground control points or previously georeferenced data, convert between raster and vector data structures, and more.

There is also an excellent QGIS plugin called SCP, the Semi Automatic Classification Plugin (mentioned above), that allows you to download and analyze satellite imagery, create color composites, and do thematic classifications to produce land use and land cover maps derived from satellite data.

You also have full access to the excellent and comprehensive GRASS, SAGA, and Orfeo systems. Access to the core raster functions is through the `Raster` menu. SCP is accessed through its own SCP menu once you install it using the *Plugin Manager*. The GRASS raster functions can be accessed either through the *Processing Toolbox* or through the GRASS QGIS interface. Shown below is the `Raster` menu:

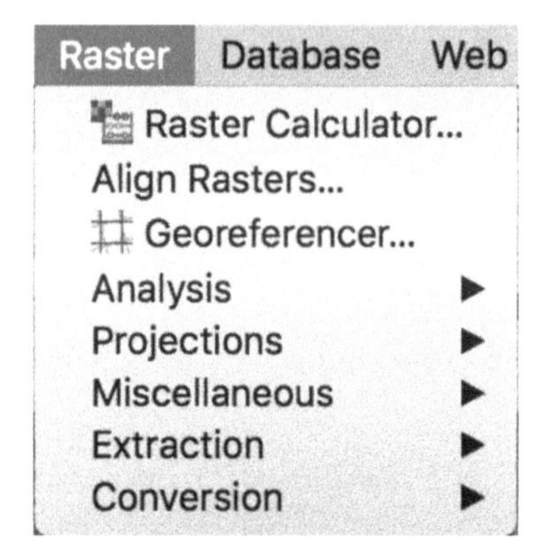

8.3 Importing and Exporting Raster Data

It is simple to both import and export raster data of a wide variety of formats using QGIS. To import a raster file, choose `Layer->Add Layer->Add Raster Layer` from the main menu:

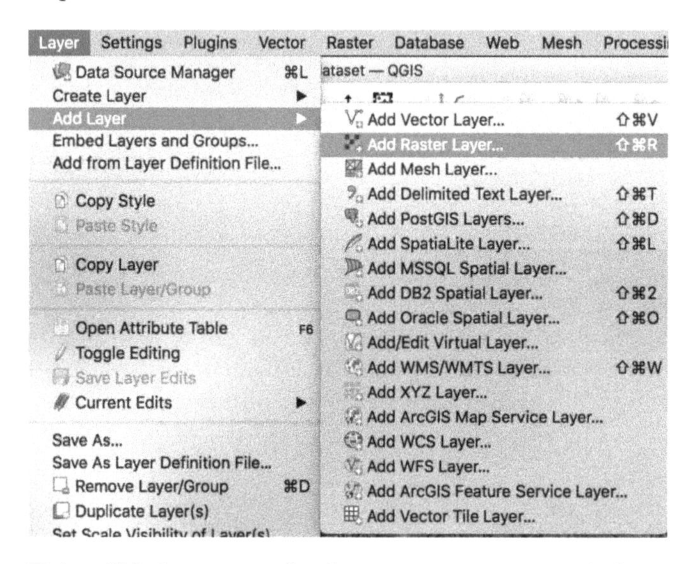

This will bring you to the data source manager window:

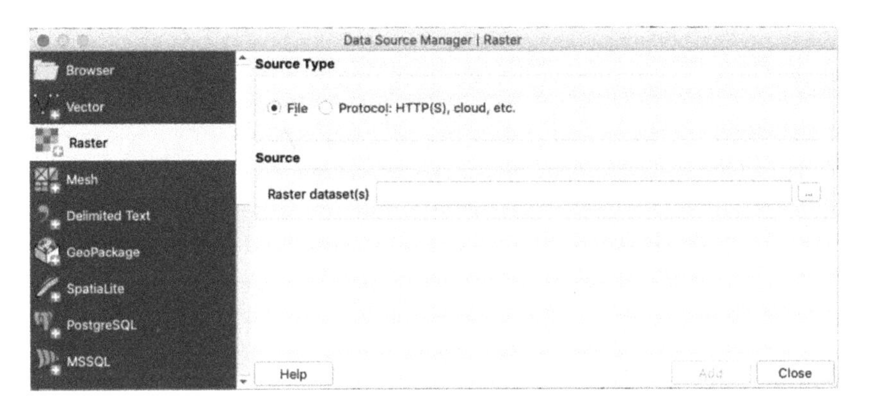

Navigate to where your raster file is stored by clicking on the three dots icon, and then click on the file filter drop-down at the bottom of the dialog to see the list of available raster data formats (the list below is only half of the available formats).

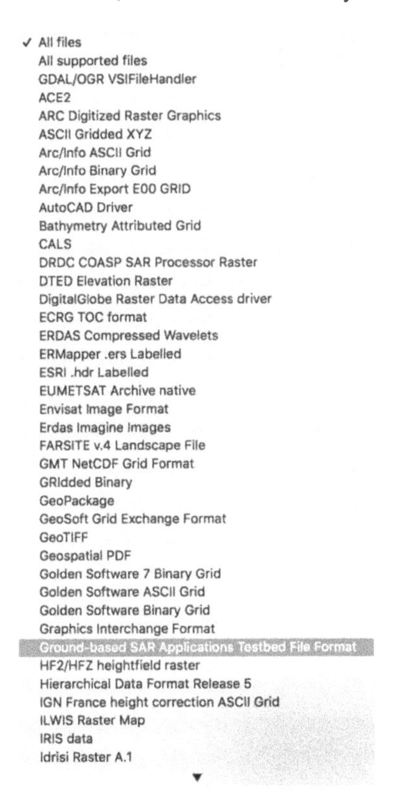

You can drag any of these files into your QGIS map area as well. If you wish to export a raster file, right-click on the raster file in the *Layers* panel and choose `Export->Save As`.

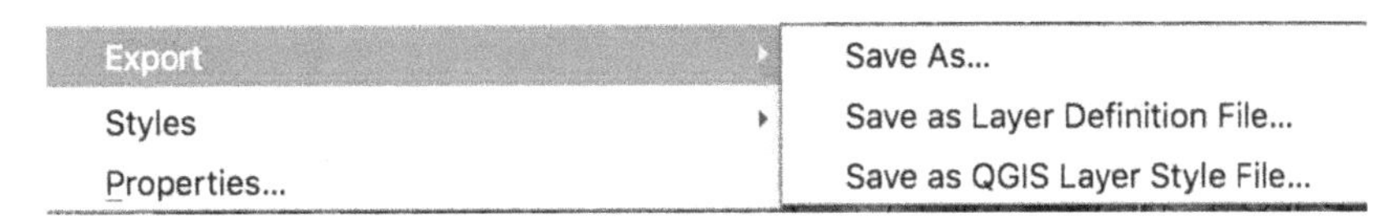

This brings up the *Save Raster Layer as* dialog:

If you click on the *Format* drop-down you will see the large number of export options that are available.

8.4 Raster Functions in QGIS

Using raster data in QGIS is simple. Looking at the `Raster` menu will show you the options:

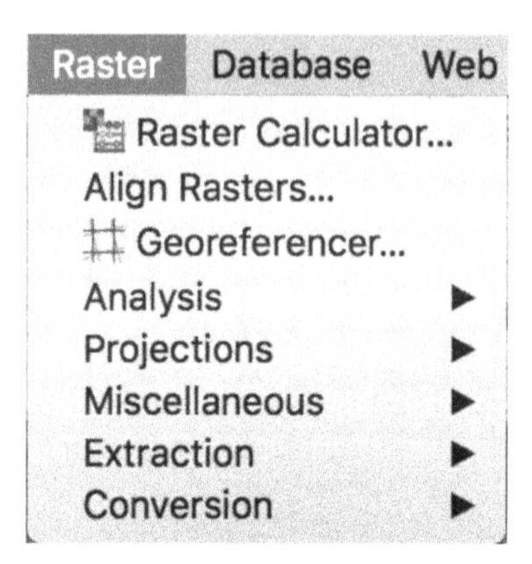

Keep in mind that these are the QGIS raster capabilities, and that there are far more functions in the *Processing* toolbox with full GRASS, SAGA, Orfeo, and other functions. Many provide similar functions, but together these provide a full range of raster GIS capabilities.

Go through the available Raster functions. You see we can do layer creation and analysis, alter projections, and extract data. The `Conversion` menu options provide conversion of raster to vector and vector to raster. We can also access the powerful `Raster Calculator` (discussed at the end of this chapter). We can georeference raster images, maps, etc. using the `Georeferencer`, which is covered in Chapter 11.

8.5 Raster Functions in the Processing Toolbox

The raster capabilities in QGIS are only a small part of our raster toolbox. In the `Processing` menu you will find many raster functions in GDAL, GRASS, and SAGA:

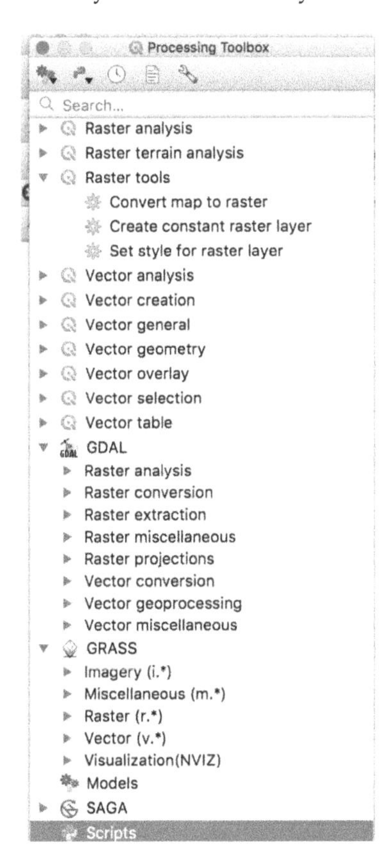

8.6 Terrain Analysis

In the *Processing Toolbox*, under *GDAL*, there are many tools that let you do analysis on raster Digital Elevation Model (DEM) data, including generating derived slope, aspect, hillshade, relief, and ruggedness index.

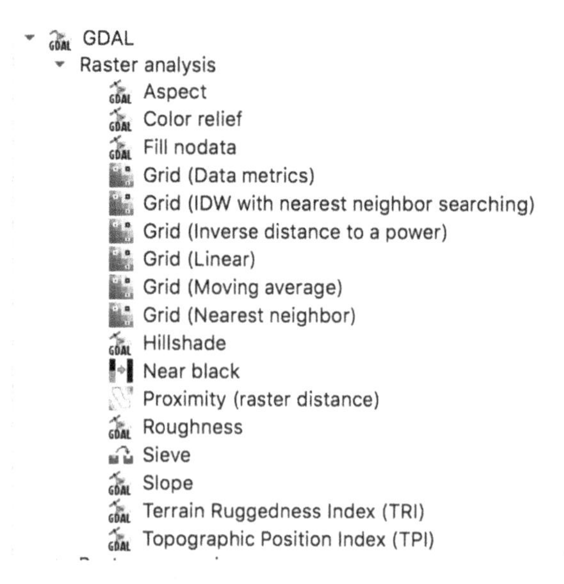

As we'll see in the next chapter, we can do these operations, and many more, using GRASS and SAGA in the *Processing Toolbox*.

8.7 Working with Digital Elevation Model (DEM) Data

Using raster elevation data is a standard part of GIS display, analysis and modeling. We can then create the derived slope, aspect, terrain roughness, and other layers, as well as do 3D perspective views, create watershed and drainage networks, etc. But first, we need a Digital Elevation Model (DEM) representing the terrain in the study area to work with. A DEM can be created from vector contours, if they are available, through the process of interpolation. Fortunately, DEM data are now readily available. NASA acquired global DEM data in 1992 using the Space Shuttle and the SIR-C radar system. (`https://www2.jpl.nasa.gov/srtm/cbanddataproducts.html`)

There are many other sources of DEM data, including very high resolution LiDAR data. For information on where to find DEM data, see:

- `https://www.usgs.gov/faqs/where-can-i-get-global-elevation-data`
- `https://opentopography.org`

These data are publicly available online and cover almost the entire world (areas near the north and south poles are not yet available). There are many data repositories for this data, and our South African study area is in tile `srtm_42_19`. I downloaded this very large .zip file and here it is, covering the entire Southeastern section of South Africa:

Drag the `srtm_42_19.tif` file onto your QGIS map area, and drag it to the bottom. Be sure to drag the .tif file. I still have the Bing satellite data from QuickMapServices, so you can see that as well as the grayscale DEM. Our study area is the small red speck:

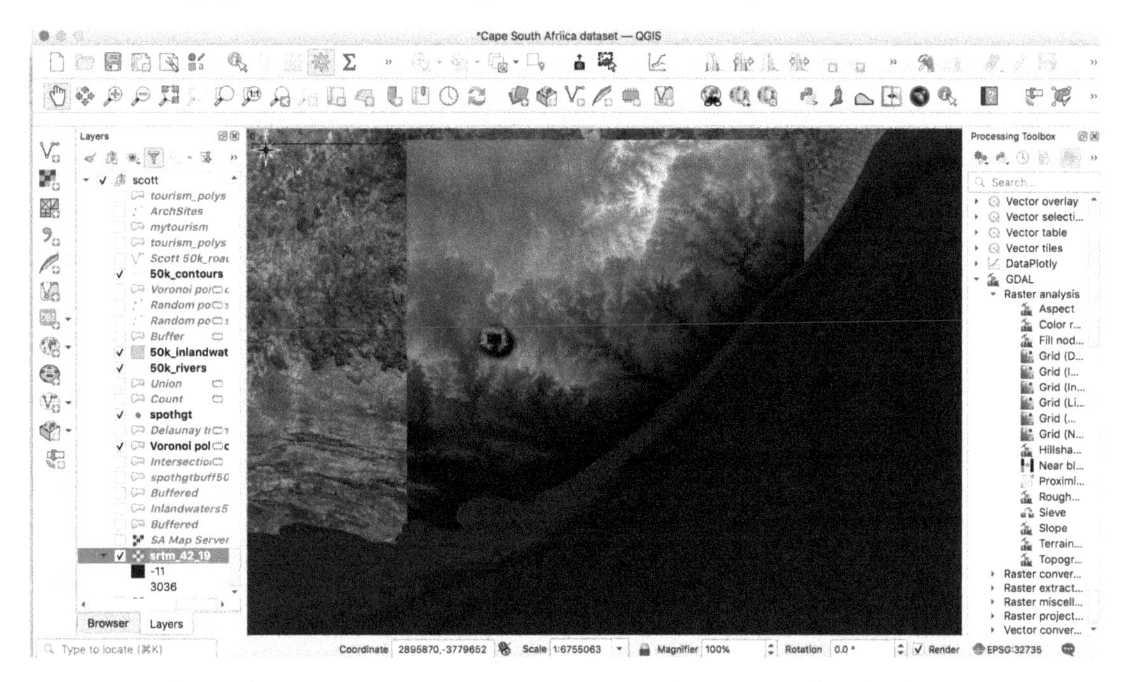

Next, we will use the `Raster->Clip Raster by Extent` function to clip this large DEM down to our study area. Choose any of the Cape GIS layers to outline to dataset area, but you could pick a geographical extent as well. Note the GDAL script at bottom, which is what is actually being processed. This is shown below:

You can now see that we have created a new raster DEM that only covers an area the size of our study area. This is a scratch (temporary) file, so be sure to save it by right-clicking on it, go to `Export->Save As` and give it a name.

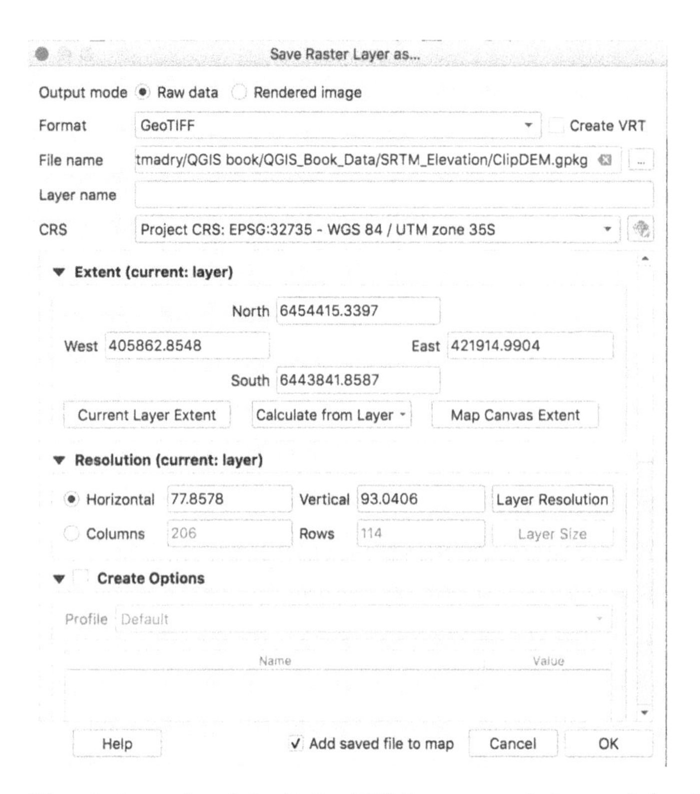

Note I chose the default GeoTiff format, and changed the CRS to our local UTM, and saved it in the directory of my Cape dataset. Click *OK* and it will create and load it. Now unclick the SRTM file and scratch file to make our new layer visible:

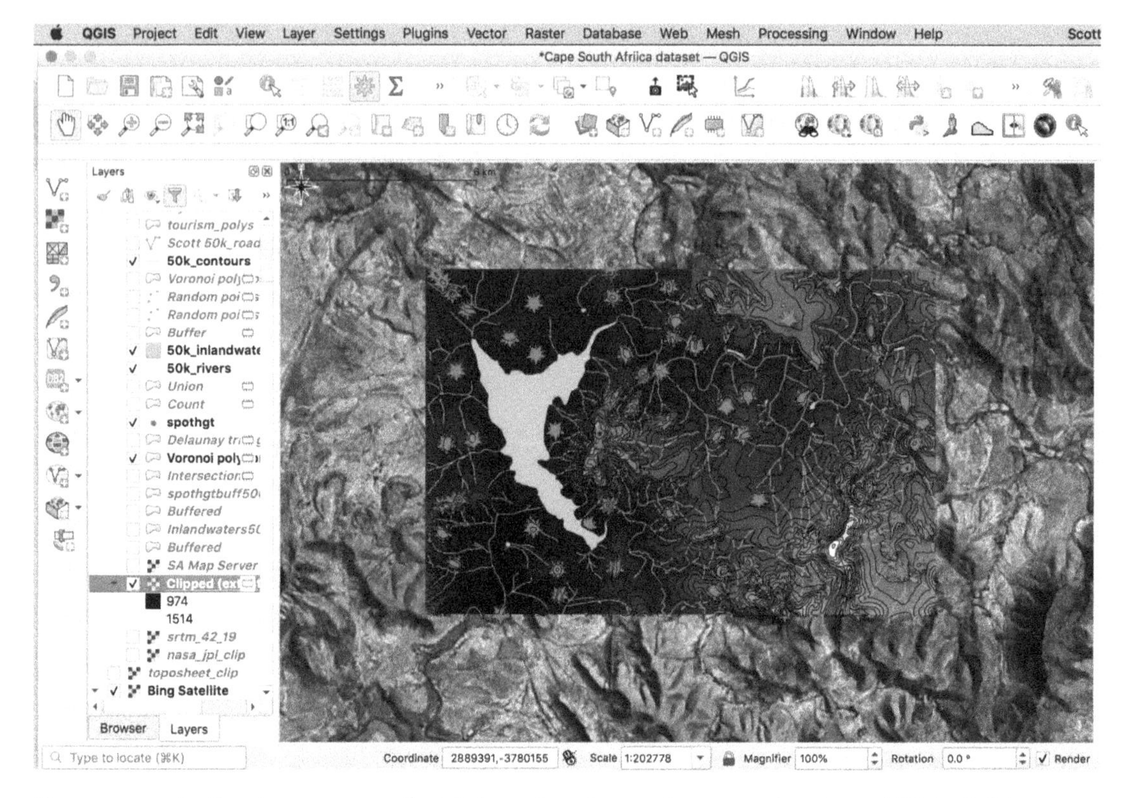

You can see in the *Layers* panel that it has elevation values ranging from 974 to 1514 meters above sea level. If you don't like the black DEM, go to `Properties->Symbology` and for *Render type*, choose *Singleband pseudocolor*, click *Classify*, then *OK*.

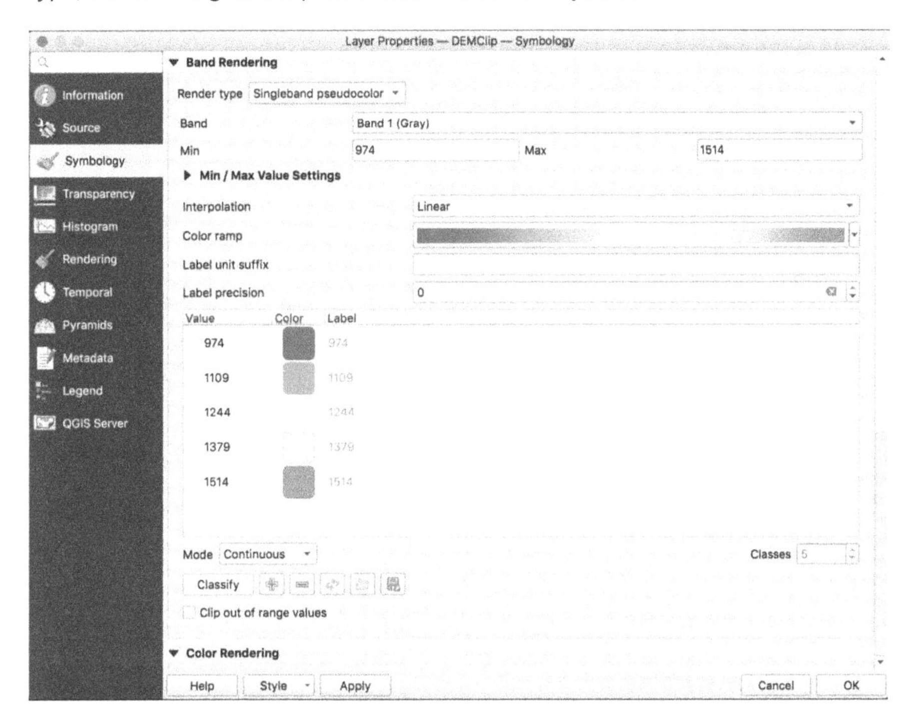

The result is shown below, with the vector contours and a few other layers overlaid:

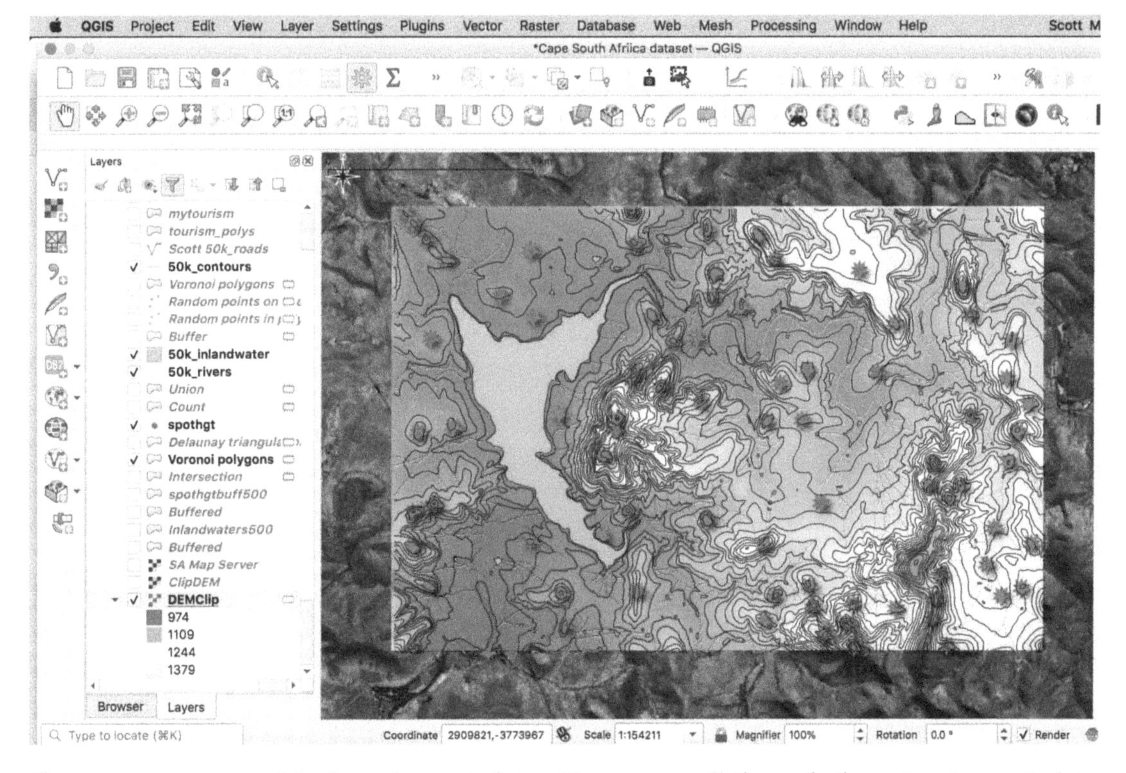

Go to `PROPERTIES` and look at the metadata. You can scroll through the extensive metadata automatically created by the GDAL script, but it's up to you to fill in the rest of the information.

Do this for every new layer you create.

Now that we have a DEM of our area, you can create the derivative terrain layers of slope, aspect, etc. First, right-click on our new `ClipDEM` layer in the *Layers* panel and `Zoom to Native Resolution` and `Zoom to Layer`. We do this because we want to be sure we are working with the best quality raster data, which is one major difference between raster and vector files. Vectors are always just vector arcs and nodes, but raster data can be displayed and processed at any cell size.

Now go to `Raster->Analysis->Aspect` and create a new aspect file, representing the compass direction that the slope of the terrain is facing:

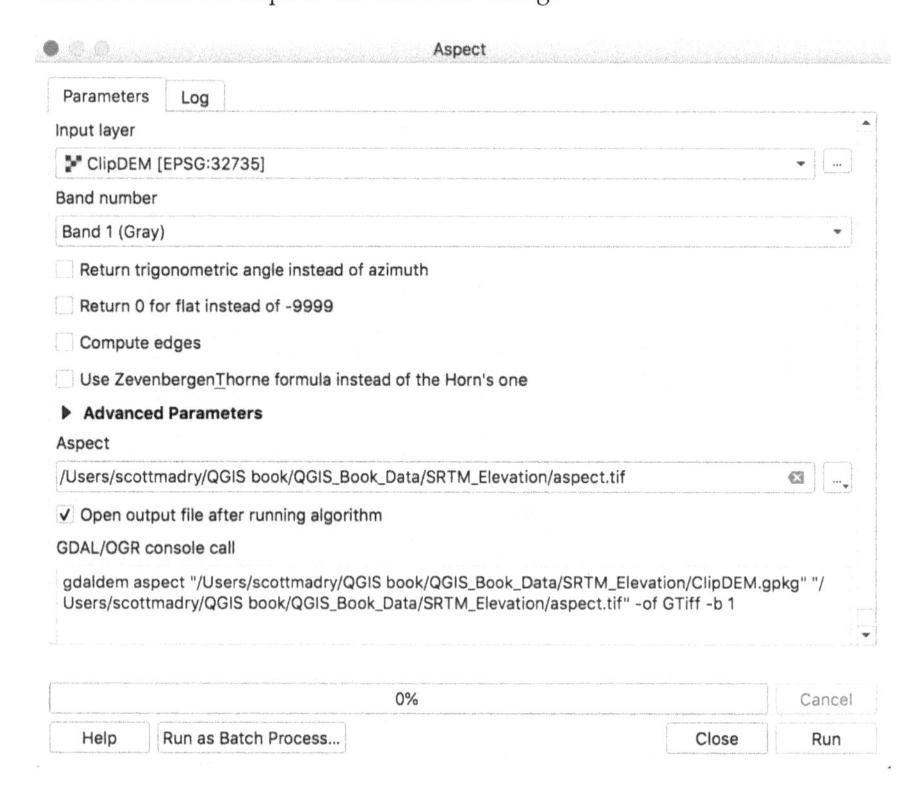

Note the options the GDAL script at the bottom that shows you what it is actually doing. Here is the resulting aspect map:

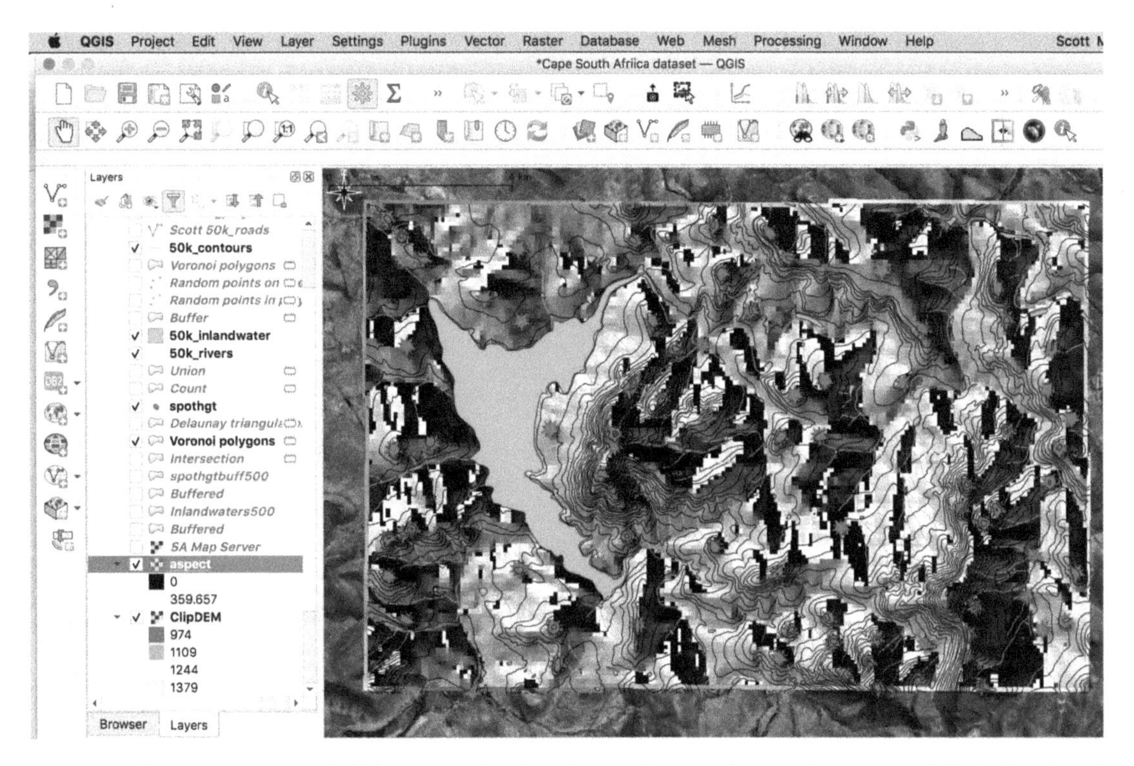

You can also create slope, hillshade, and other derived raster layers from your DEM. Go ahead and do these—it is quick and easy for a small area like our study area. We can also do line-of-sight analysis, showing where we can see from a given location, trafficability studies, showing the least cost path from one location to another, ecological and vegetation studies, 3D visualizations, and a wide variety of hydrological studies such as creating a drainage network and catchment basins. These all begin with a good DEM. Also note that GRASS and SAGA have a wide variety of these tools, all found in the *Processing Toolbox*.

The SRTM data is quite coarse for our small study area, but you can create DEMs from point or vector contour elevation data, if available. A new method is by using airborne drones and LiDAR data, which is extremely high resolution data, and there are QGIS plugins for managing UAV drone aircraft to acquire LiDAR and other data. GRASS has a complete LiDAR suite of capabilities to go from a raw point cloud to a finished DEM.

For processing LiDAR data using QGIS and GRASS, see the excellent QGIS/GRASS LiDAR tutorial:

- `https://courses.neteler.org/importing-and-visualizing-las-lidar-files-in-grass-gis-7-r-in-lidar/`
- `https://www.pointsnorthgis.ca/blog/blog/dem-from-lidar-using-open-source-software-tutorial/`

8.8 Interpolating Raster DEMs from Point Data

If we have either point or vector contour data, we can convert these to a raster DEM, and then all of the various derived products. In the `Cape_dataUTM` folder of your Cape database there is a point shapefile named `elevation_p1000`. Load this, and you will see that it is a point dataset of 1,000 elevation points, covering a small area in the southeast of our study area.

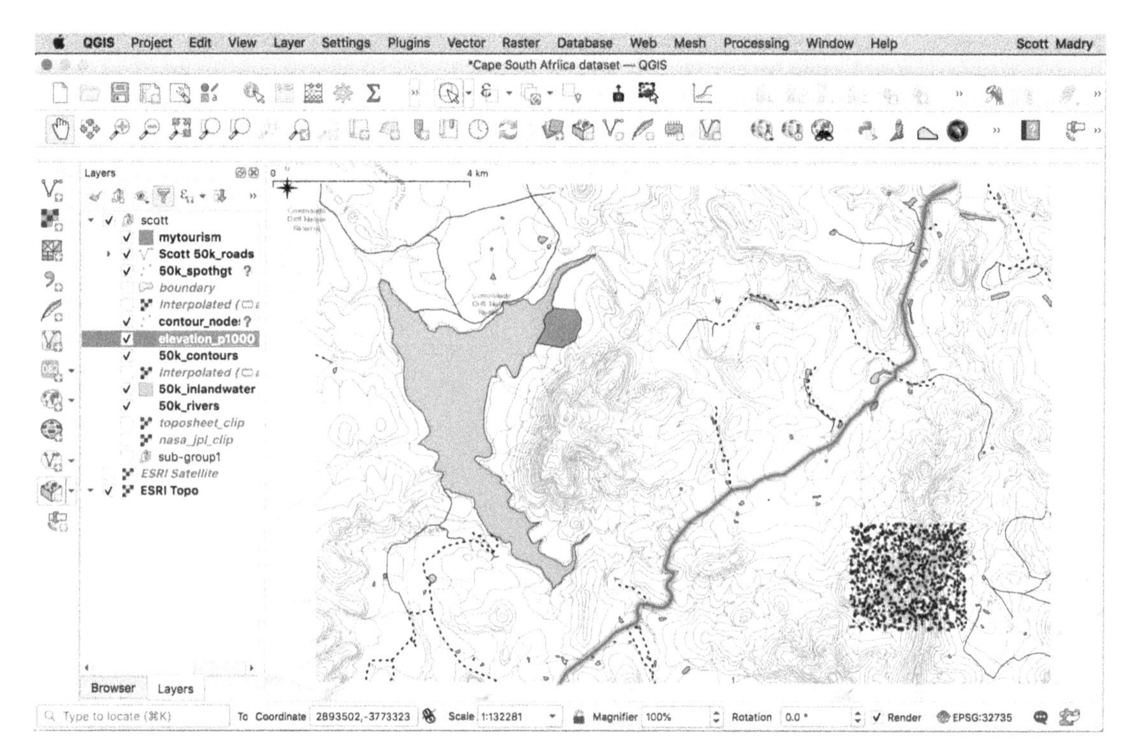

Right-click on the `elevation_p1000` layer, and open the attribute table You will see that it has a number and the elevation in the `value` field:

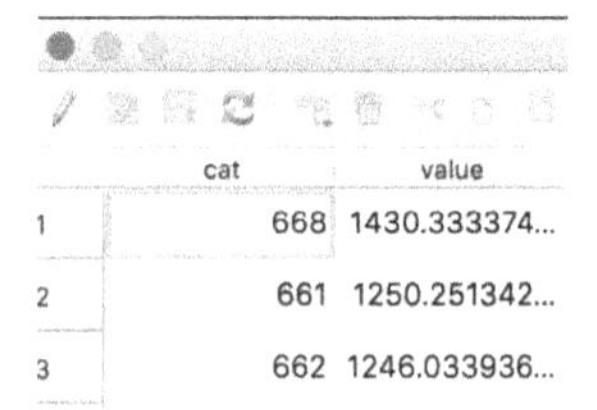

	cat	value
1	668	1430.333374...
2	661	1250.251342...
3	662	1246.033936...

To understand our data better, click on the Sigma icon $\sum$ and let's look at the data:

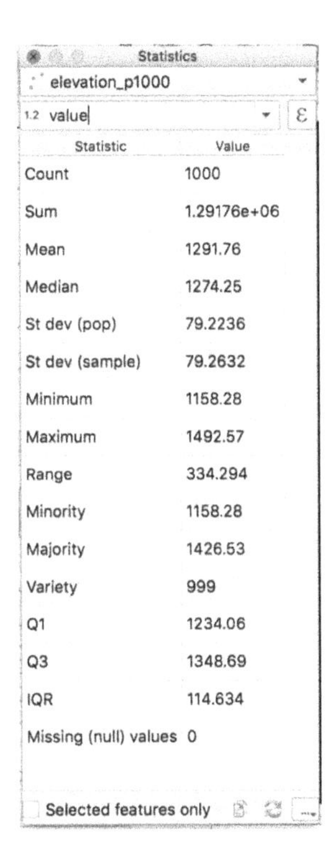

Let's first change the display of our points, so go to `Properties->Symbology`, and change the display from *Single Symbol* to *Graduated*, pick a color ramp, and make ten classes, as shown below:

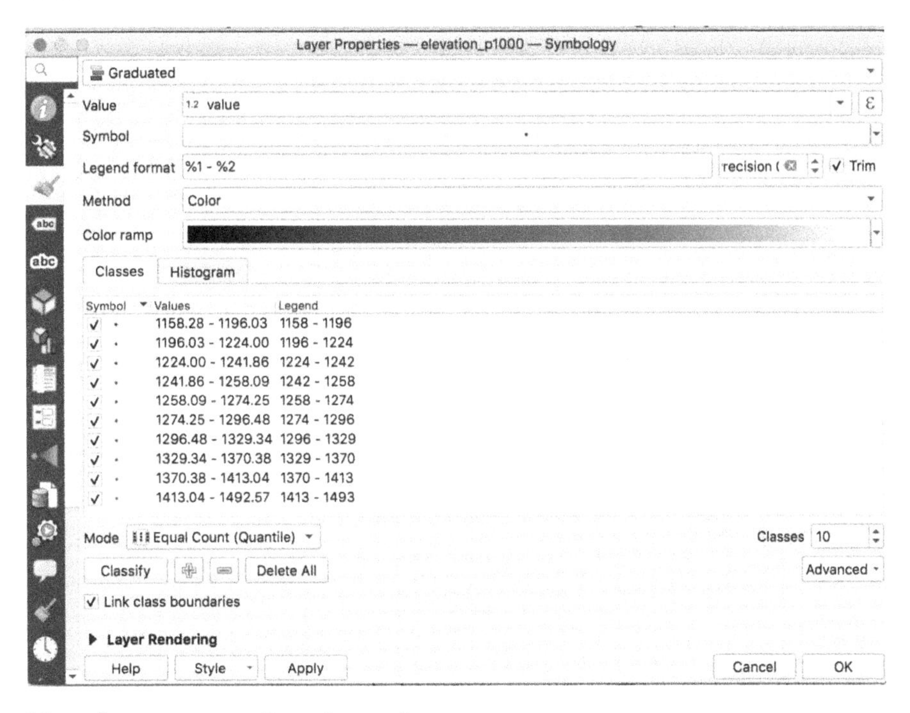

Note that you can adjust the mode:

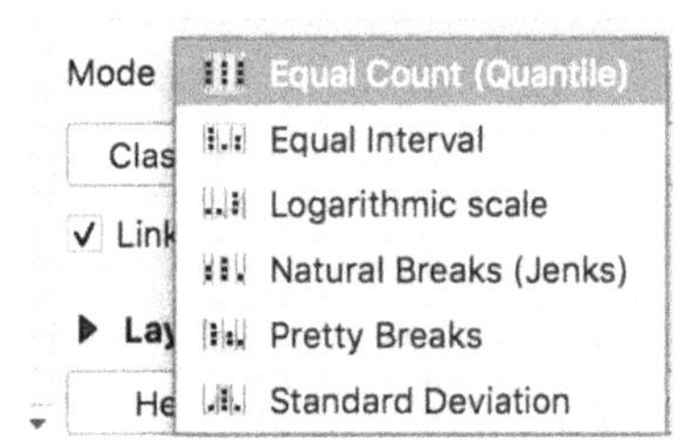

Click on the *Histogram* tab and *Load Values*, and it will show you an interesting representation of the distribution of the ten classes.

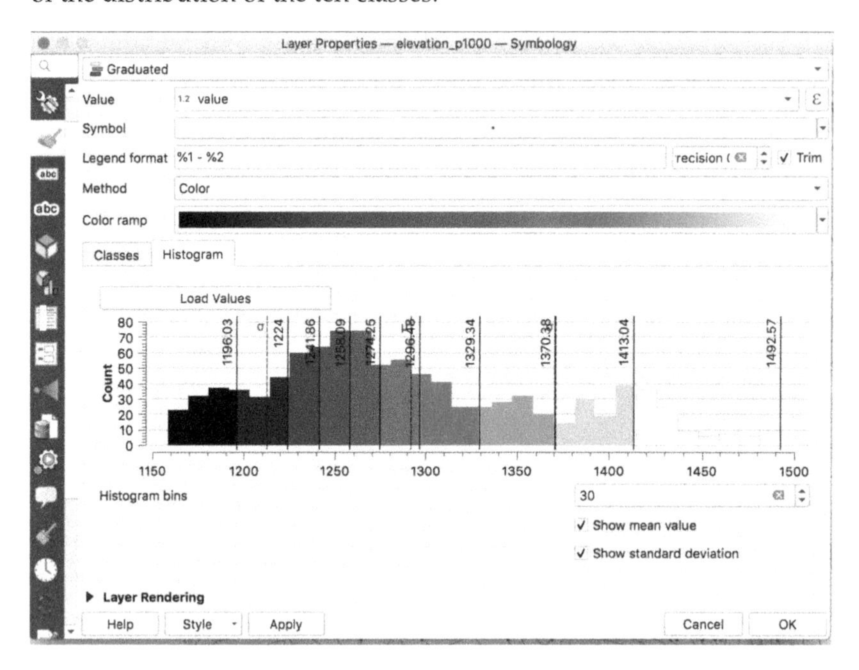

Click on the *Show mean value* and *Show standard deviation* boxes at lower right to see these. Now zoom in to just that area, either by `Zoom to Layer Extent`, or with your mouse, and look at the vector contours, points, and a background map. Here I'm using the ESRI Topo background map from the QuickMapServices plugin.

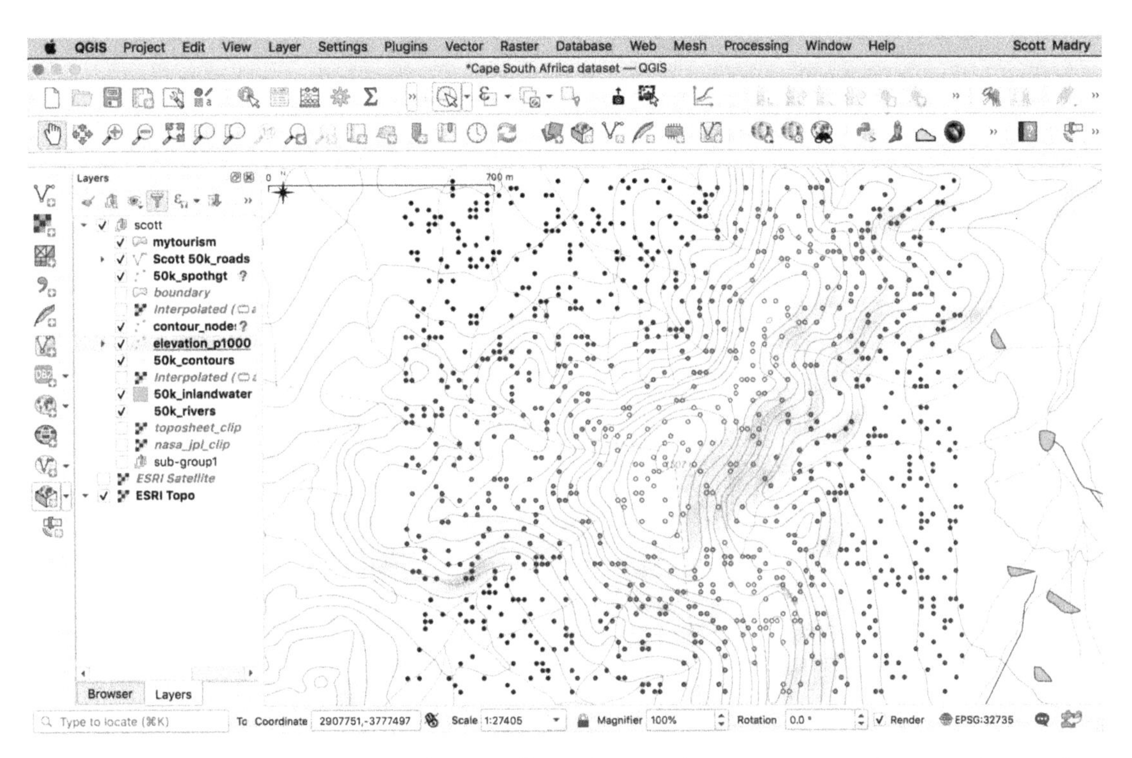

Now choose `Raster->Analysis` and you will see four modules starting with `Grid` to interpolate these point files into a raster DEM using different techniques.

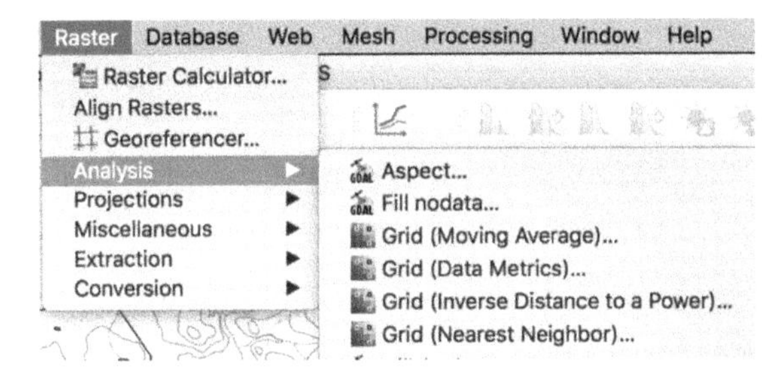

Lets try the `Grid (Nearest Neighbor)` module. Choose your point file, set a search radius, and be sure to set the *Z value from field* to the `value` field in your attribute table. Click *Run*.

Here is the result, with the vector contours and points above it:

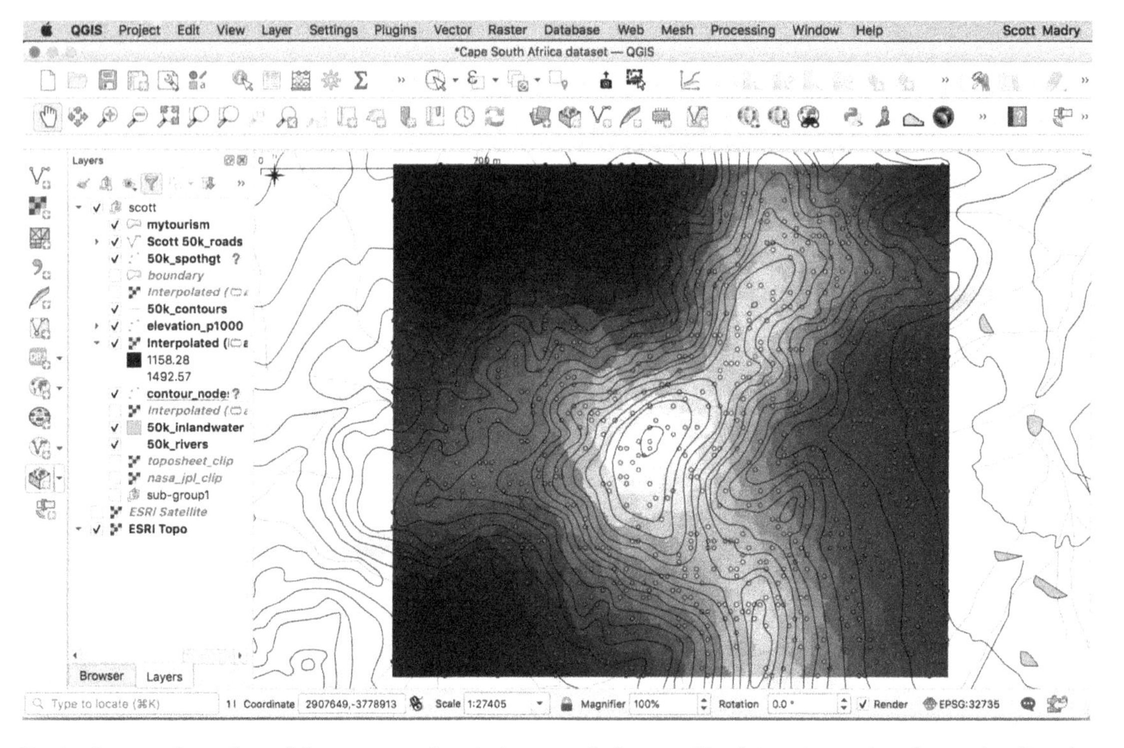

Let's change the color table, so go to *Symbology*, and choose *Singleband pseudocolor* as the *Render*

type, pick a color ramp, click on *Classify*, and then *OK*

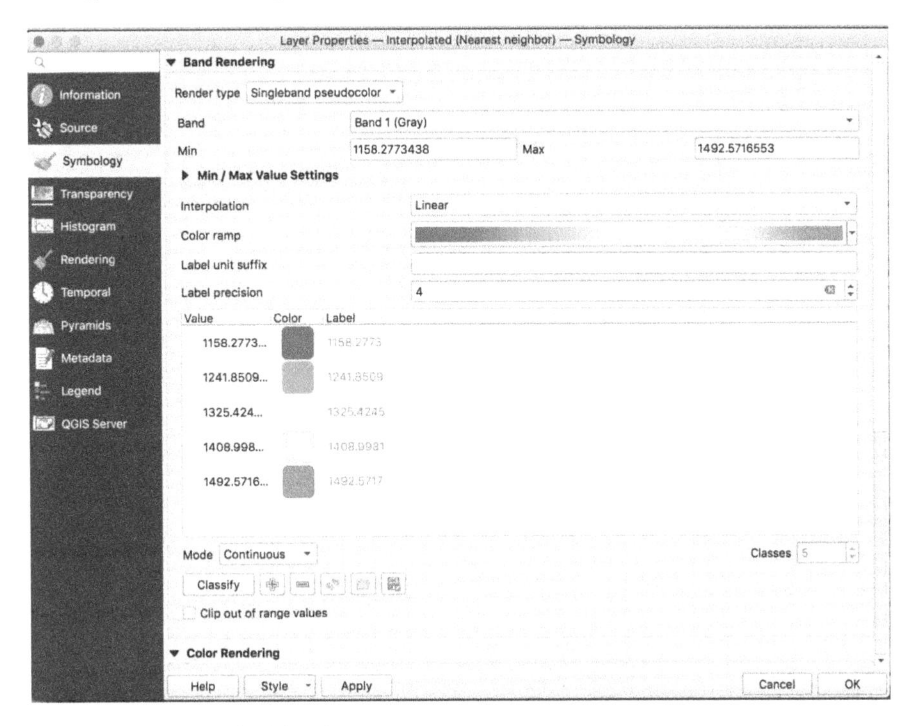

Our map now looks like this:

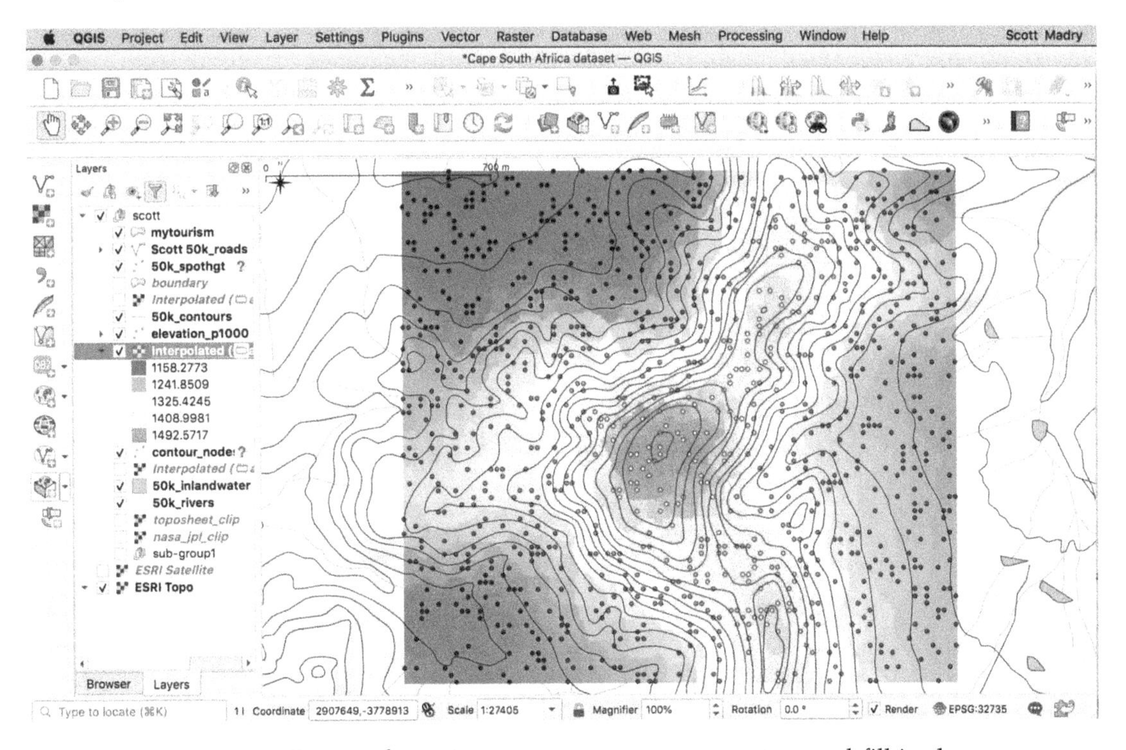

Now open up a new 3D view, by going to `View->New 3D Map View`, and fill in the parameters by clicking on the wrench at the top right.

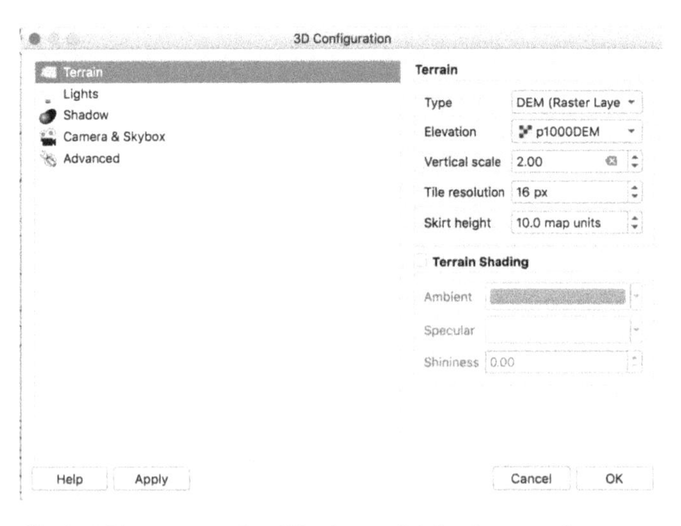

Click *OK* to create the 3D view, which gives us this:

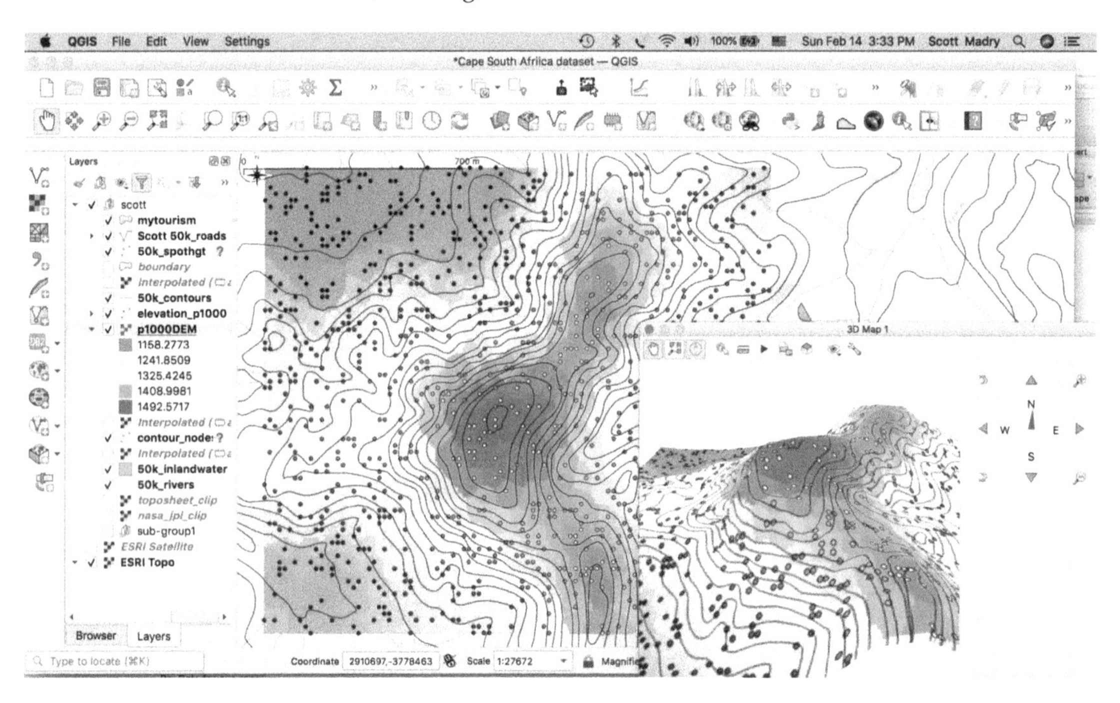

I reversed the color table to highlight the mountain. Now you can overlay other data, rotate, zoom in, make and save a fly through, etc. Don't tell me this is not fun—and really simple to learn. QGIS rocks.

Each of the various `Grid` options in the `Raster->Analysis` menu give slightly different results, based on the inputs and calculation used. Note that you can also do this in GRASS and in SAGA as well. You can create a point file from the vector contours to do this, and once you have a DEM you can create all the derived slope, aspect, watershed, cost surfaces, etc. Knowing the difference in the methods and parameters is important to create accurate data. Remember "garbage in, garbage out", so read the documentation and learn what you are actually doing when you push that *OK* button.

8.9 The Raster Calculator

The *Raster Calculator* is a very powerful analytic engine for processing raster data. It allows you to implement raster map calculator analysis on your files for a wide range of functions. This is actually how the various raster analysis processing for deriving slope and aspect from elevation data (and others) is done. Most remote sensing data processing is done using the *Raster Calculator*. It is the Swiss Army knife of raster GIS and something you need to learn about.

Raster data are just a grid square, where each square has a given numerical value. These may be the elevation, slope, or the reflectance value of a channel of a remote sensing imagery. Raster data is just a two-dimensional raster array, to use the computer science language.

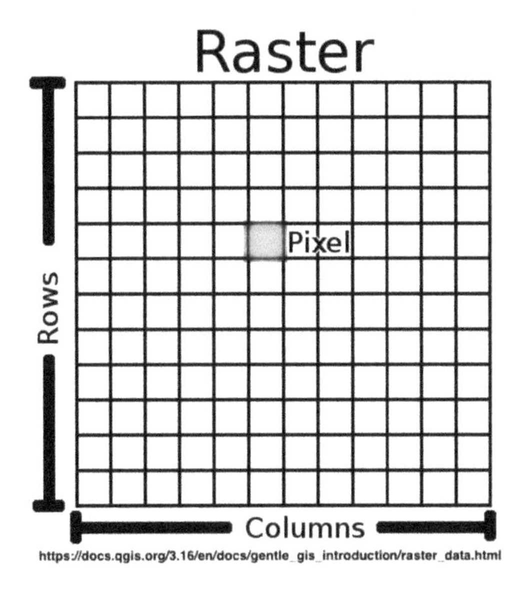

If we zoom way in to a raster image, we see that it is just a grid of cells. This is the spatial resolution of the image, and it can be altered easily. With the Raster Calculator, we can create slope and aspect from a DEM, do satellite image processing, and create raster masks, filters, along with many more useful operations. Go to the `Raster` menu and click on the `Raster Calculator`:

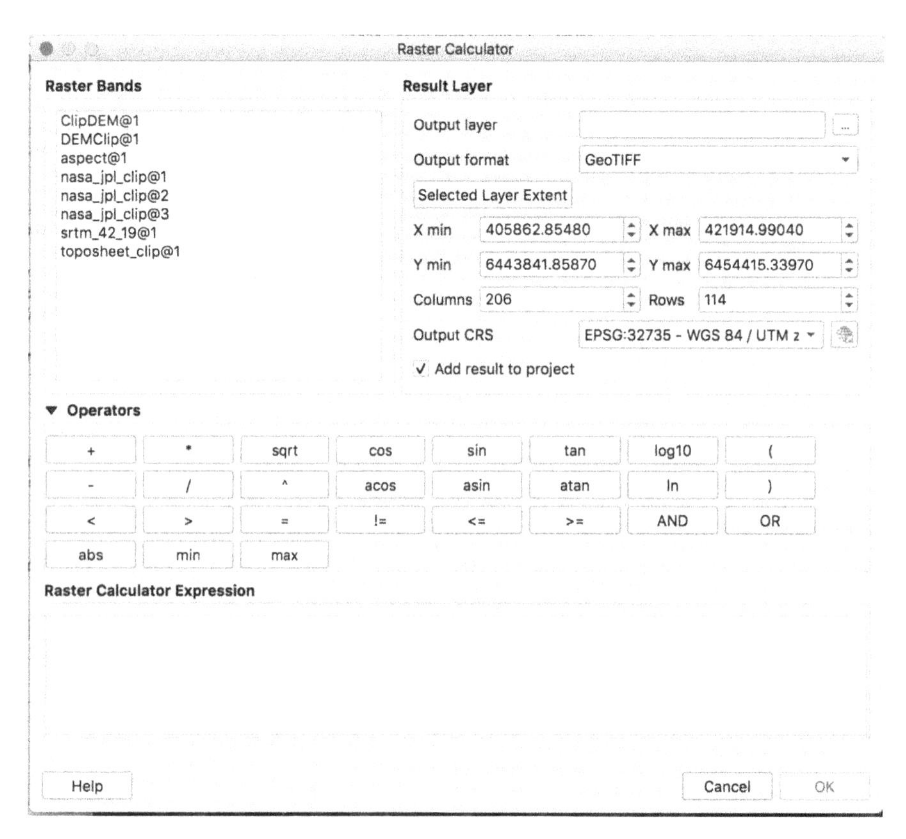

Under *Raster Bands* are several raster layers we have loaded in QGIS.

The simplest example of how to use the raster calculator is to convert our `ClipDEM` from meters to feet. In the raster calculator, double-click on *ClipDEM* and it will appear in the *Raster Calculator Expresssion* box. Click on the * operator and type *3.28*. Enter a name for the resulting file—call it `DEMFeet`. Note at the bottom left it says this is a valid expression, and click *OK*.

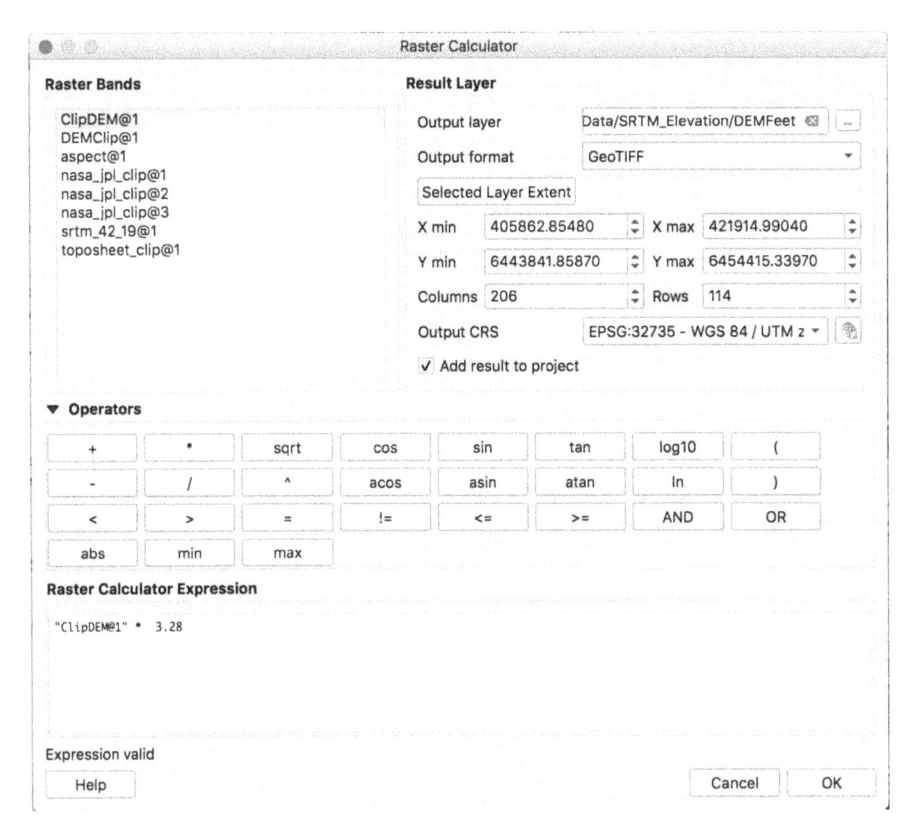

Look in your Layers Panel and QGIS display, and you have now created a new DEM where each cell value is the product of its original value multiplied by 3.28. The DEM now ranges from 3197.72 to 4965.92 feet. This is a most simple example—expressions can be very complex but, in the end, we are simply doing math on the value of each grid cell and creating a new file with the resulting values. Look again at the mathematical operators that are available:

▼ Operators

+	*	sqrt	cos	sin	tan	log10	(
-	/	^	acos	asin	atan	ln	)
<	>	=	!=	<=	>=	AND	OR
abs	min	max					

Learning how to use the *Raster Calculator* is one of the most important GIS skills you can acquire. It is an extremely powerful and useful capability. You can create masks, reclassify data, do image processing, and much more, all using the *Raster Calculator*. To learn more, see the QGIS manual: `https://docs.qgis.org/3.16/en/docs/user_manual/working_with_raster/raster_analysis.html` and the GRASS module `r.mapcalc` manual: `https://grass.osgeo.org/grass79/manuals/r.mapcalc.html`

As an example, what if you want to know (as an archaeologist, or biologist, or environmental planner) where in your study area is every location that is above a given elevation, below another elevation, facing southeast, with less than 15% slope, on one of three soil types, with pine forest land cover, within 200 meters of a stream, and greater than 500 meters from a road? Boom—you can create a new layer with the *Raster Calculator* that meets all of these criteria, and faster than it takes for me to type this. Create a point file in the centroid of each area, and download this to your GPS, and go and visit each of these sites. The raster calculator is the possibly the most powerful tool in the GIS toolbox.

Years ago, in the early GRASS GIS days, we had a saying that if you could not figure out how to do something in the GIS: "*When in doubt, r.mapcalc*". You can do an amazing range of things with the *Raster Calculator*, so take the time to learn how to use it.

9. The Processing Toolbox, Modeler, and Python Console

QGIS is only one of several open source GIS systems, and, in the past, it was a difficult and complicated process to use these capabilities in any sort of a structured way. Each application has their own user interface, and often used their own data formats, requiring conversion of data.

In QGIS 3.0 and up, there is a major new analytic component that has been added with the incorporation of the SEXTANTE project and tools into the core QGIS capabilities. SEXTANTE is now found under `Processing` menu, which organizes the analysis functions of several other GIS systems into one place, including SAGA, ORFEO, GRASS, and the native QGIS tools. All of these can now be executed from the QGIS interface and the *Modeler*, as well as in batch processing mode. You can create your own scripts or models that create defined workflows and that allows you to put together commonly used sets of routines that can be saved and reused. This is somewhat similar to the ESRI Arc toolbox and Arc Model Builder.

These, and many other spatial data processing tools, are now available directly from the `Processing` menu. The greatest benefit of this approach is that we now have access to many hundreds of vector, raster, point, remote sensing, modeling, and other tools, and it handles all of the data format issues internally. This has made a massive increase in the power and usability of QGIS, and it is an example of the willingness of the open source community to work together for the common good. This is the future direction of QGIS, and I think we will see the processing tab replace the vector and raster menus in the future.

When you click on the `Processing` menu, you see:

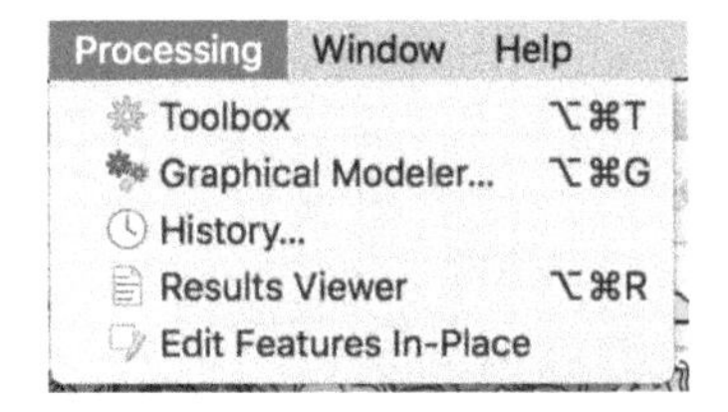

Click on the `Toolbox` and you'll see a list of all the available processing algorithms, divided into logical groups. The *Toolbox* panel, shown below, gives you access to all of the various processing tools in QGIS, as well as the other tools, including GDAL, GRASS, SAGA, models, scripts, etc.

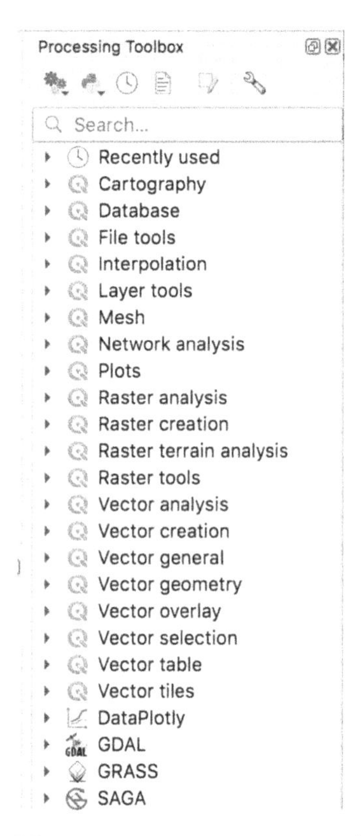

Hover over any icon to see what that module does:

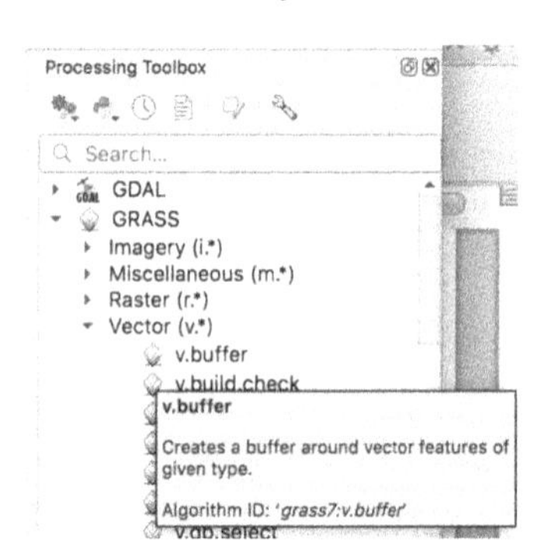

At the top are several icons that provide you with access to creating new models or opening models, scripts, history, and results.

The first icon, Models allows you to create a new model, open an existing model, or add a model to the toolbox. If you choose to create a new model it will open the *Model Designer* dialog:

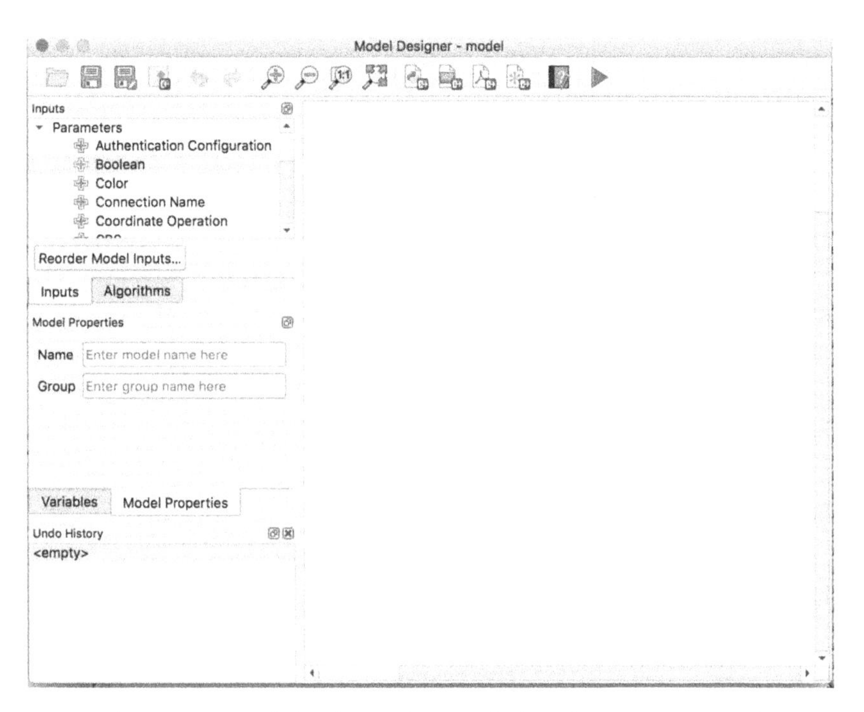

Along the top you will find a toolbar with familiar icons to work with:

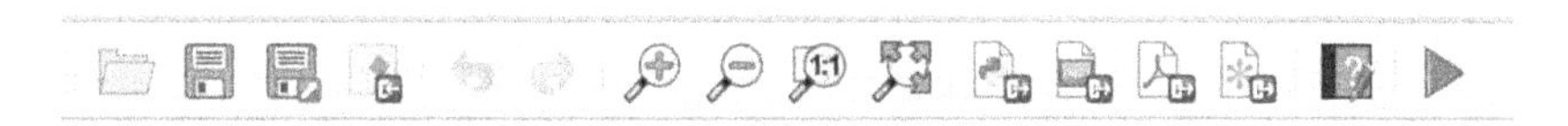

In the *Processing Toolbox*, the second icon is for scripts and lets you create, open, and add scripts to the toolbox:

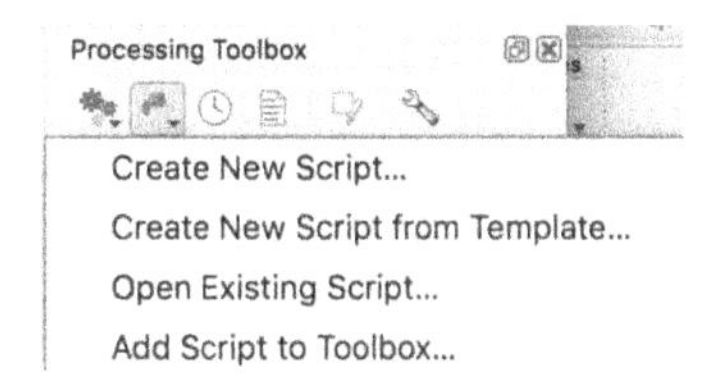

You can enter any term in the search line to search for available functions in the toolbox. You will also note that at the top are shown the most recent modules you have used. The history icon shows you a log of all the processes that you have run in this work session:

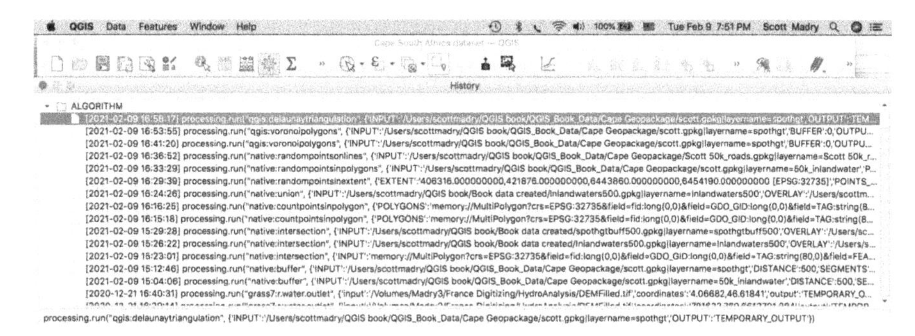

You can double-click on an entry in the recent list and it will return you to that particular QGIS module and process—very useful.

If you click on the right-most icon it takes you to the processing options page:

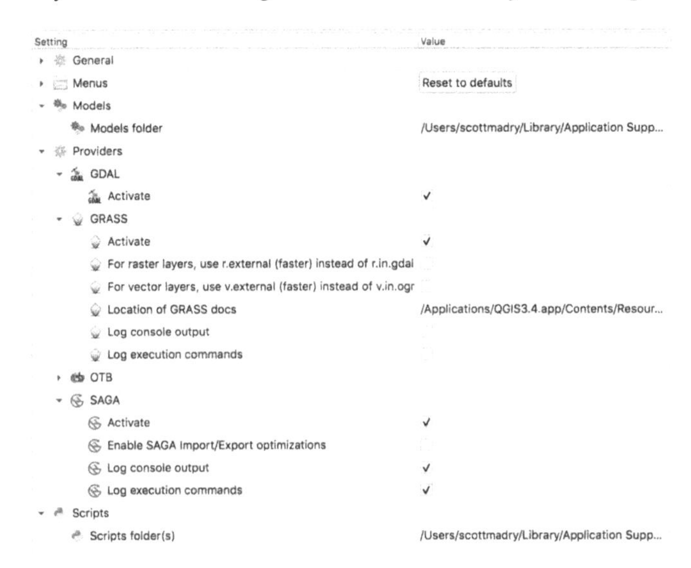

This shows you your processing tools, models, scripts, etc, and you can define parameters.

The *Processing Toolbox* is where the actual tools are located, and it organizes all of the native QGIS functions into logical groups, which can be accessed by clicking on the small triangle at left. For example, here are the QGIS *Vector analysis* functions:

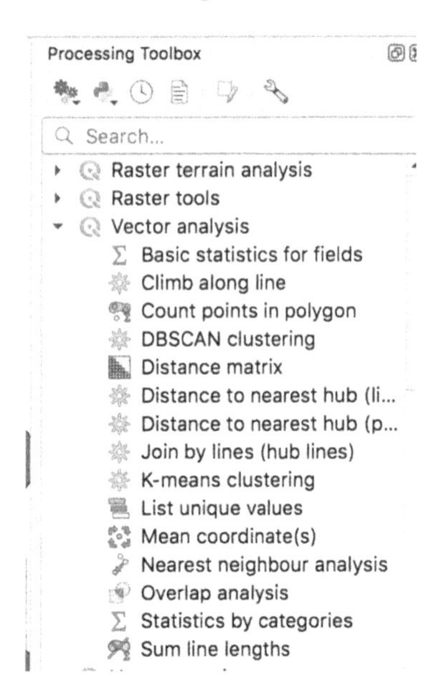

Double-clicking on any individual module will open the dialog box to run the module. For example, double-clicking on *Count Points in Polygon* will open this:

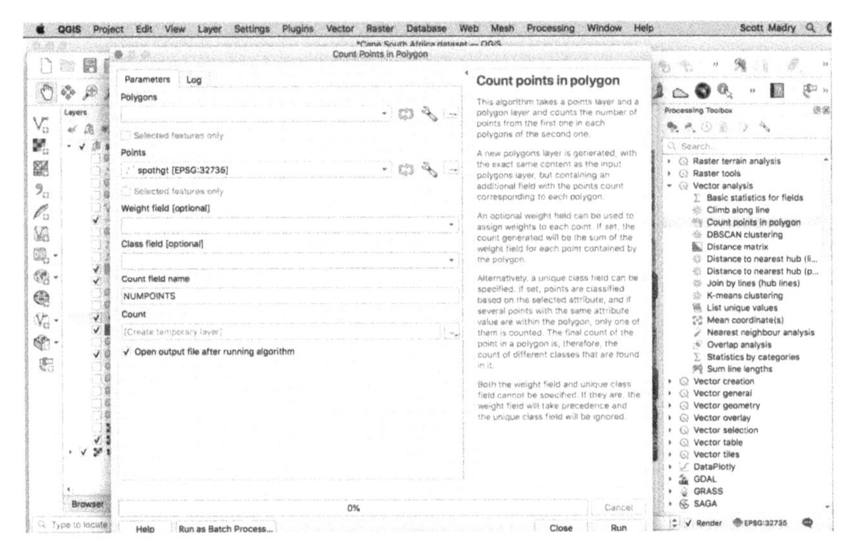

All of the QGIS analysis functions can now be accessed through the toolbox.

9.1 Using GRASS, GDAL, the Orfeo Toolbox, and SAGA tools within QGIS

There are many open source GIS tools, and the *Processing Toolbox* allows us to access not only QGIS native capabilities, but also these other, powerful toolkits, including GRASS, GDAL, SAGA, and ORFEO. One of the most useful and powerful features of the *Processing Toolbox* is that it brings all of these tools together under the QGIS interface, and it automatically does the data prep, reformatting, and conversion between all the various file formats for you.

The result—all the GRASS GIS raster and vector analysis functions (over 400) are now available directly in QGIS without converting your shapefiles or Geodatabase files into and out of the GRASS format, which is unique to GRASS.

For example, if you want to use a GRASS function on a QGIS vector shapefile, the toolbox will automatically covert the QGIS vector shapefile (or geodatabases or raster GeoTIFs) into the needed GRASS data structure, run the analysis, reconvert the file format back to a shapefile, and put the new file into your QGIS project workspace and display. It will then delete all the temporary files.

This brings the full analytic capability of GRASS, and all the other systems, into QGIS without worrying about file formats, conversions, and different user interfaces. You can run all individual modules directly from the *Processing Toolbox*.

GDAL The Geographic Data Abstraction Library

GDAL is a translator library for vector and raster geospatial data formats that is released by the Open Source Geospatial Foundation OSGeo (`https://gdal.org`). This library comes with a raster and vector data model, as well as useful utilities for data processing and translation. QGIS uses many GDAL routines, but now you can run them directly using the *Processing Toolbox*. Click on the *GDAL* tab and see all of the available tools. For example, click on *Raster analysis* to see the tools available:

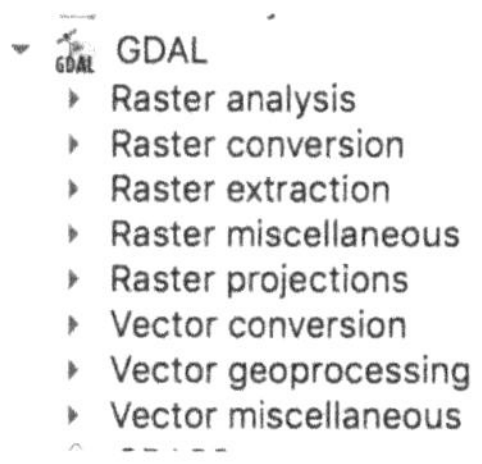

- GDAL
  - Raster analysis
  - Raster conversion
  - Raster extraction
  - Raster miscellaneous
  - Raster projections
  - Vector conversion
  - Vector geoprocessing
  - Vector miscellaneous

Double-click on any one to see and run the individual module. It is the same with all of the GRASS and SAGA modules. There are now hundreds of vector and raster modules for you to use as integrated functions within the QGIS user interface.

SAGA GIS The System for Automated Geoscientific Analyses

The SAGA GIS was first developed in 2004, and is an open source GIS that was originally developed by the University of Göttengen in Germany, in their Department of Physical Geography (`http://www.saga-gis.org/en`). It has extensive vector, raster, remote sensing, modeling, geostatistics, terrain analysis, and modeling of dynamic simulations.

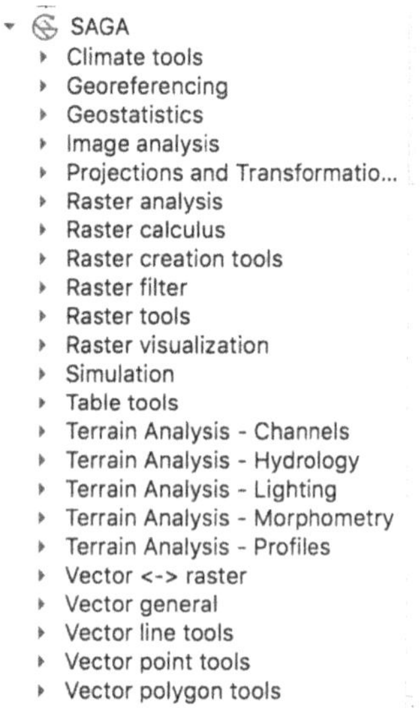

- SAGA
  - Climate tools
  - Georeferencing
  - Geostatistics
  - Image analysis
  - Projections and Transformatio...
  - Raster analysis
  - Raster calculus
  - Raster creation tools
  - Raster filter
  - Raster tools
  - Raster visualization
  - Simulation
  - Table tools
  - Terrain Analysis - Channels
  - Terrain Analysis - Hydrology
  - Terrain Analysis - Lighting
  - Terrain Analysis - Morphometry
  - Terrain Analysis - Profiles
  - Vector <-> raster
  - Vector general
  - Vector line tools
  - Vector point tools
  - Vector polygon tools

OTB- The Orfeo Toolbox

The Orfeo Toolbox was developed by CNES, the French national space agency, for the analysis and processing of space remote sensing data generated by the French Cosmo-Skymed and Pleiades satellite systems. It was begun in 2006, and is released under an Open Source license. It focuses on satellite image processing, and is very robust. Note that since QGIS 3.8, the OTB plugin is disabled, and you must install OTB (`https://orfeo-toolbox.org/download/`), then enable the plugin using the *Processing Toolbox* options icon (discussed above).

GRASS GIS

The GRASS GIS, the Geographical Analysis Support System (`https://grass.osgeo.org`) is the granddaddy of all open source GIS and is still, in many ways, the most complete and capable. It was first developed by the U.S. Army Corps of Engineers Construction Engineering Research Laboratories (USA-CERL), in 1982. It was developed to meet U.S. federal land management needs, including environmental compliance for both ecological and cultural resource management on federal lands, and for conducting environmental impact statements.

In 1997 development was taken over by Baylor University in Texas, and then in 2001 it moved to Italy. It contains over 400 modules for vector, raster, imagery, database, modeling, and even includes a true 3D voxel data structure. Grass is still a major stand-alone and very popular GIS, but it uses its own file formats, which limited interoperability, until the development of the QGIS *Processing Toolbox* fully integrated it within QGIS.

Note that GRASS modules begin with a single (or double) letter and a dot, with r. modules being raster, v. being vector, i. for imagery files (remote sensing imagery data as opposed to raster GIS layers), etc. Click on the GRASS raster modules and look at the enormous number of these.

Lets use a GRASS vector tool v.buffer, which is the GRASS vector buffer function. In the *Processing Toolbox* click on `GRASS` and `Vector`, and the module `v.buffer`:

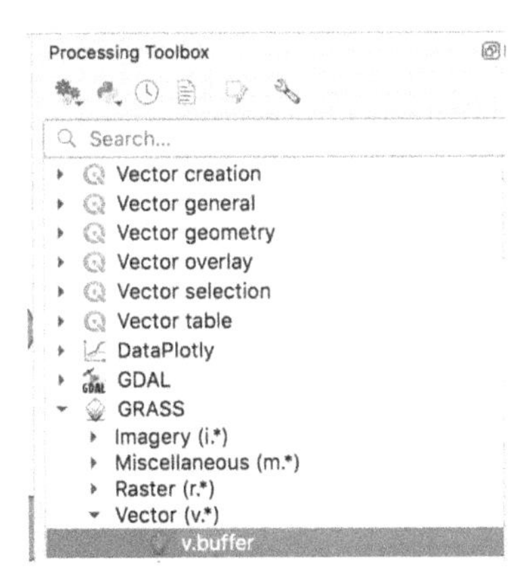

Fill out the *v.buffer* options with a 500 meter buffer distance, as shown below for our spot height points, and hit *Run*.

Here is the result, with a 500 meter buffer around our spot elevation points:

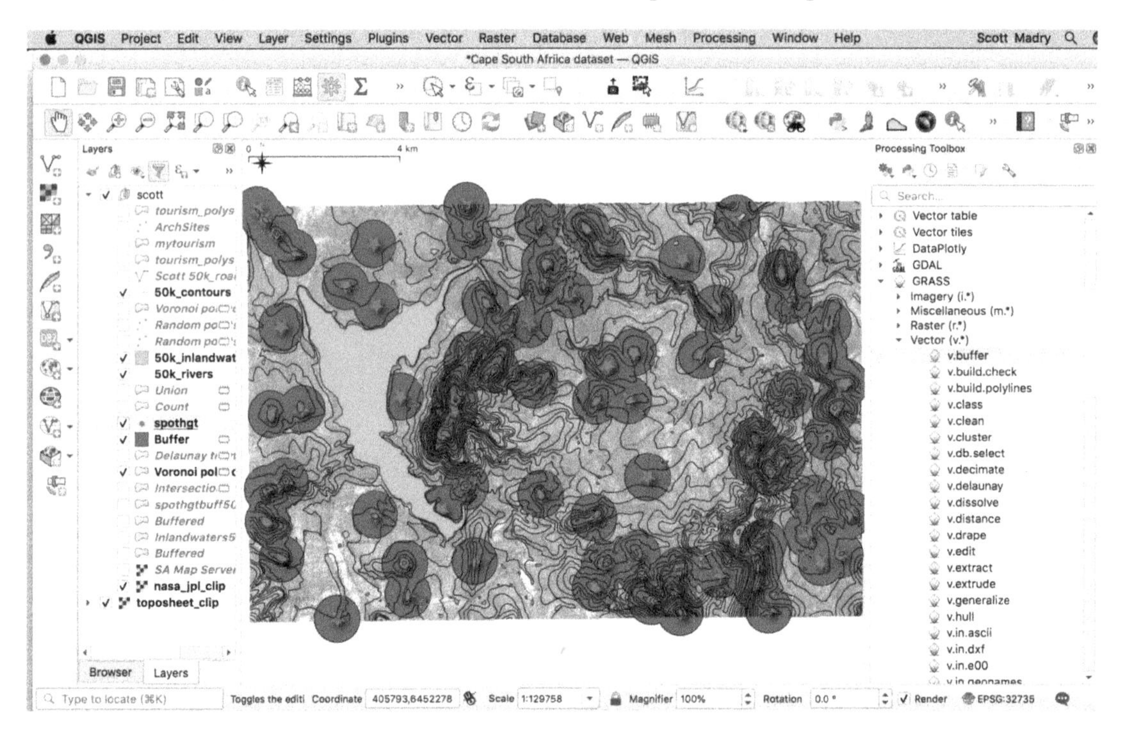

Note that this has created a scratch (temporary) layer named `Buffer`, which must be converted to a permanent file if you want to keep it. If you quit QGIS without doing this, the layer will be lost. QGIS will warn you that you have unsaved temporary layers when you quit. In the dialog box you can choose to specify where to save the buffer instead of creating a temporary file.

Using *v.buffer*, create a buffer for for `50k_inlandwater`. Remember you can go to the *Recently used* listing at the top of the *Processing Toolbox* if you want to do another *v.buffer*.

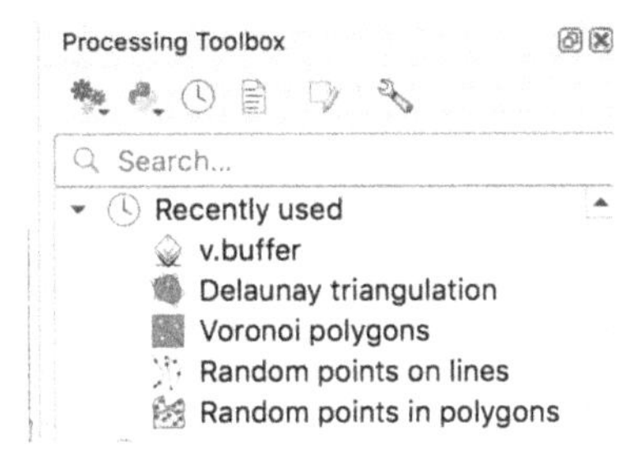

If you click on the *Log* tab you can see what GRASS has actually done to create the buffer:

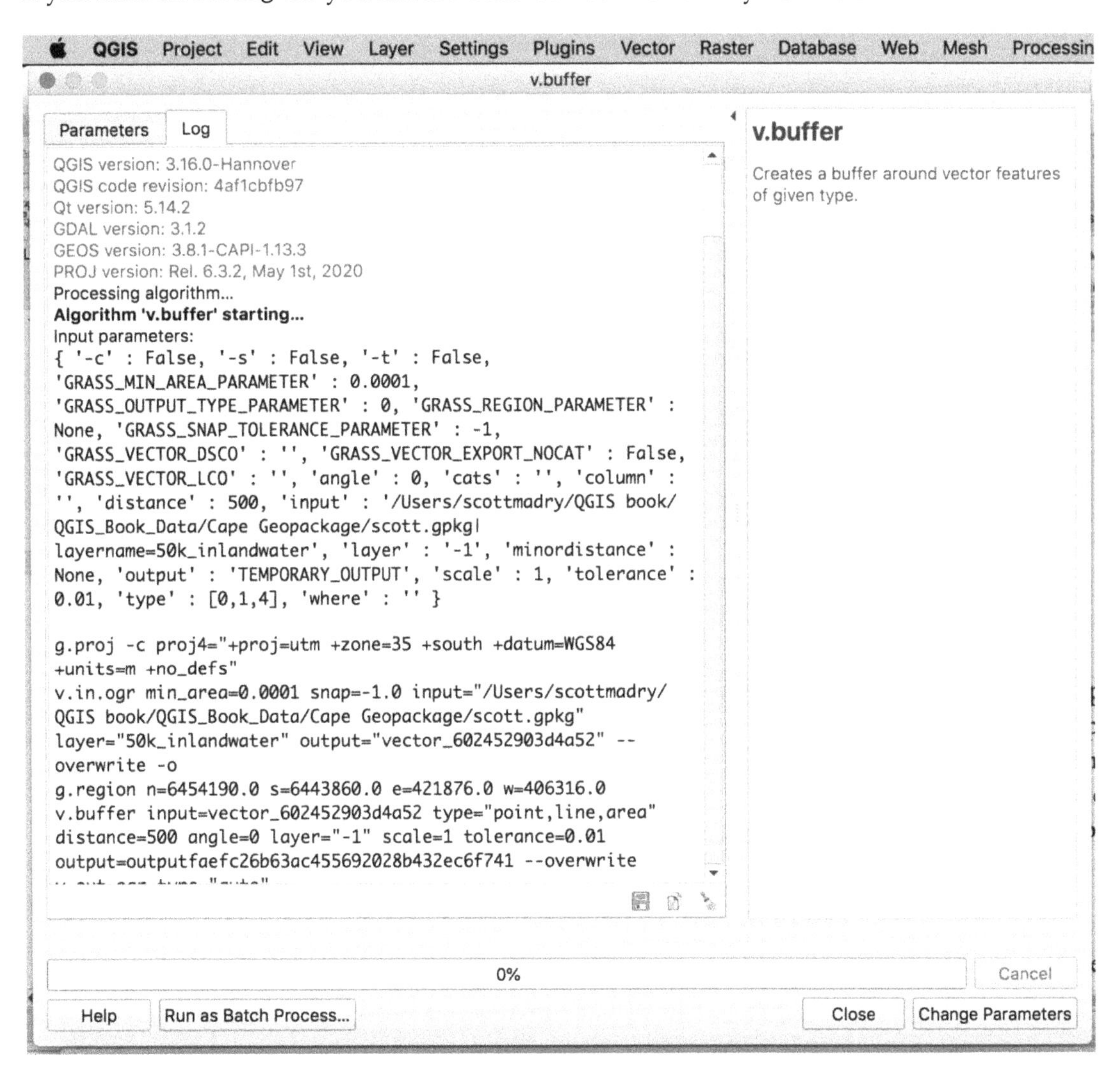

So now we have the very interesting situation where we have multiple options for common GIS functions like vector buffers, terrain analysis, and data conversions. They all pretty much do the same thing, but there may be differences. The *Processing Toolbox*, and the incorporation of these other capabilities, makes QGIS both a very powerful and yet very easy to learn and use GIS. For more information, see the QGIS manual for the *Toolbox*: `https://docs.qgis.org/3.16/en/docs/user_manual/processing/toolbox.html`

9.2 The Graphical Processing Modeler

Under the `Processing` menu you'll find the *Graphical Modeler* that allows you to create flow diagrams of complex processes that can then be run, nested, and saved for future use. This is an excellent capability for complex processes and recurring use.

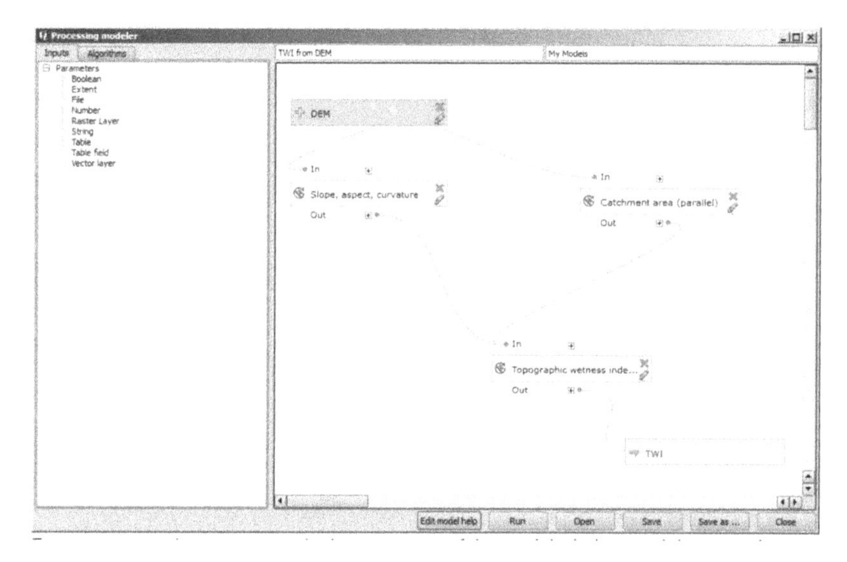

The modeler has its own working canvas and is opened from the `Processing` menu.

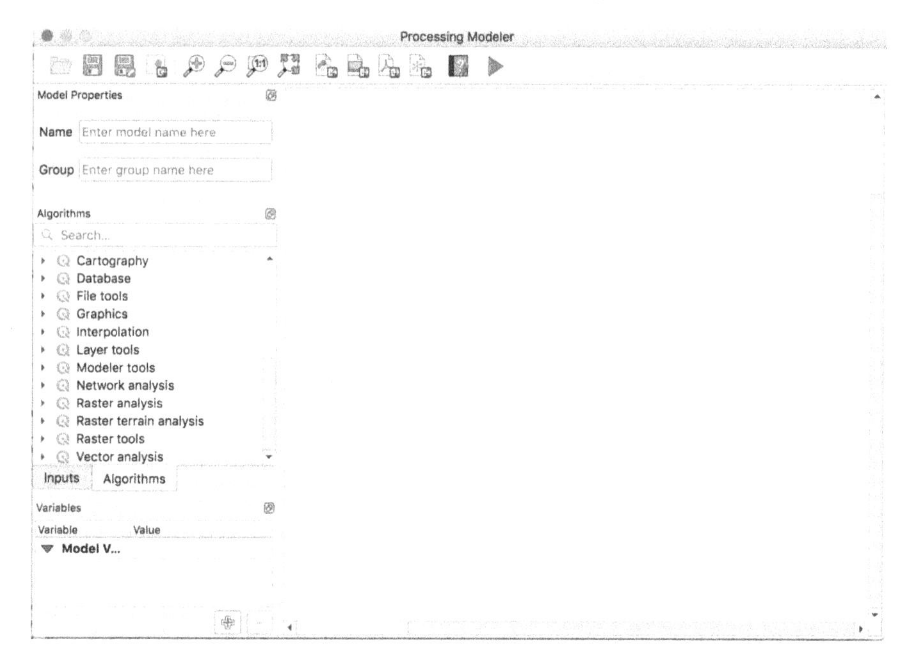

There are two windows, the available parameters and algorithms at left, and the graphical modeler window at right. At the top is a text search box, so you can just type in 'vector buffer' or 'clip' and it will show you the available tools.

There are also icons at top left. At bottom left you can choose between inputs (such as raster and vector files) and algorithms such as buffers. There are two modes, simplified and advanced, and you add an algorithm from the list at left, define its processing and the data layer to be used, and then link it to another, etc. These can be saved and run repeatedly, and can be exported as a Python script. You can run through an entire exercise using the excellent QGIS online documentation:

`https://docs.qgis.org/3.16/en/docs/training_manual/processing/modeler_twi.html`

9.3 The Python Console

There is a Python console available for scripting in QGIS. It is opened from the `Plugins->Python Console` menu as shown below.

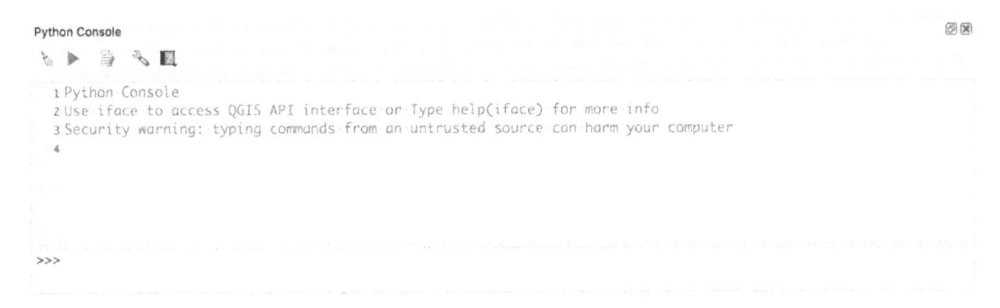

The icons at top let you clear the console, run scripts, show the editor, and get additional help.

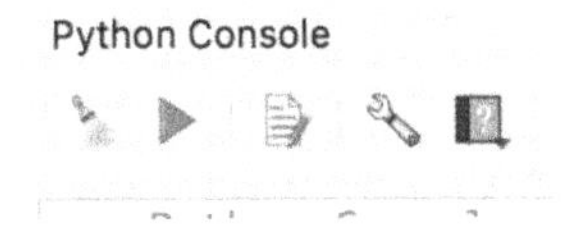

The Python console is great for doing simple one-off tasks where a plugin would be overkill. Sometimes you might want to automate a task with a small script. In addition, the console is a great way to play around with the API to learn more about how to use Python with QGIS. When writing a plugin, the console can be used to test individual bits of code to ensure it does what you want. For a more in-depth discussion of using the Python console, see *The PyQGIS Programmer's Guide* at `https://locatepress.com/ppg3`.

In conclusion, the QGIS Processing Toolbox, Graphical Process Modeler, and Python Console provide robust tools capable of enabling powerful analysis for many applications.

10. 3D QGIS

QGIS 3.0 brings a new, 3D capability to QGIS. This allows you to view your vector and raster data from a 3D perspective. Now that we have accessed and processed a DEM, as described in the Raster Data and Analysis chapter, on page 149, we can view our data in 3D.

10.1 The QGIS 3D Map Viewer

To begin, choose `View->New 3D Map View` from the main menu:

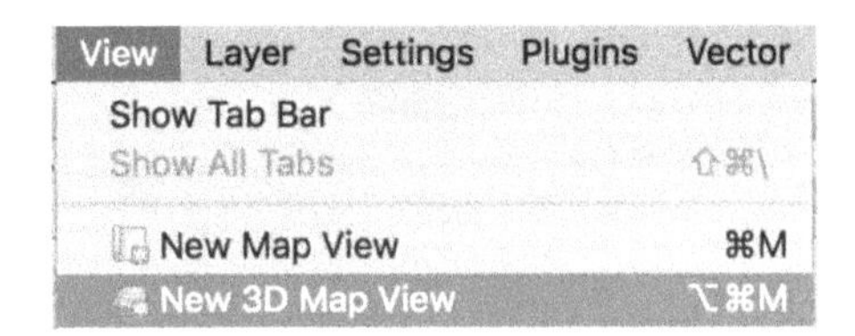

You will see a new *3D Map* window appear, that mirrors your map. You can resize it as you wish the 3D window as desired.

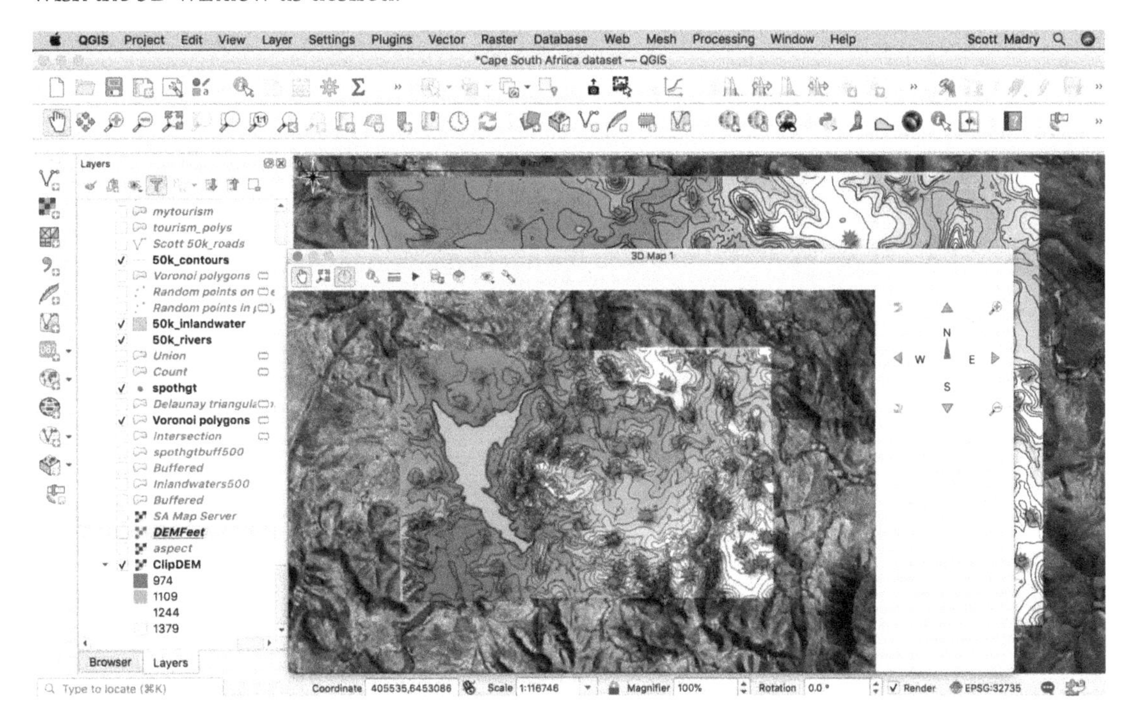

10.2 3D Viewer Icons

There are a number of icons in the toolbar at the top of the *3D Map* window:

You can hover over them to see what they do, but many should be familiar by now. Click on the wrench icon to open the *3D Configuration* dialog. After you select *DEM (Raster Layer)* for *Type*, you will choose the name of the *Elevation* file, vertical exaggeration (useful in very flat areas), tile resolution and more.

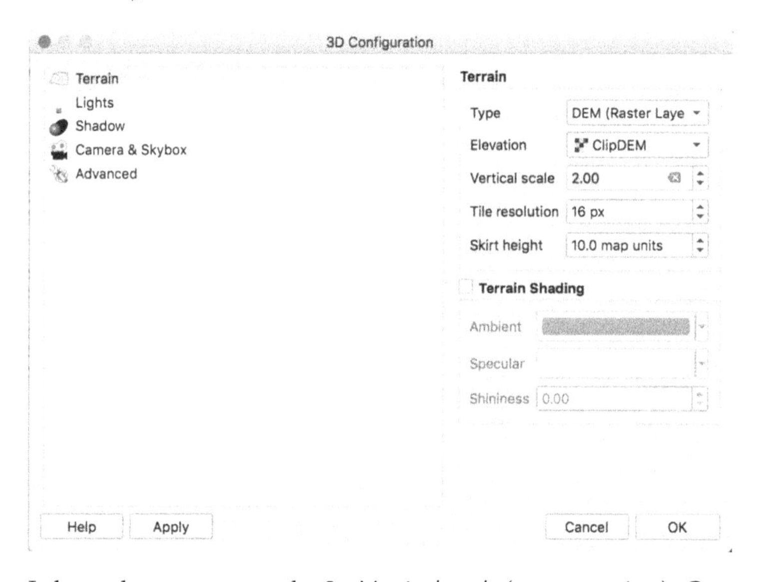

I chose the `ClipDEM`, and a 2x *Vertical scale* (exaggeration). Once you choose the DEM and click *OK*, you will see your data in 3D. It will display all vector files you currently have active. You can zoom and pan with the mouse and shift-click to rotate and see the area in 3D, or use the icons at right.

The view will display all of the data, raster and vector, that is currently visible in your main QGIS window. At first your image appears to be flat, but you can rotate the image using the red compass arrow. To see in 3D, put your mouse in the middle of the image, and click `shift` and drag your mouse or track pad down, and you will see the image lay down in 3D. You can zoom in and out by right clicking and moving your mouse, or use the magnifying glass icons at top right and left.

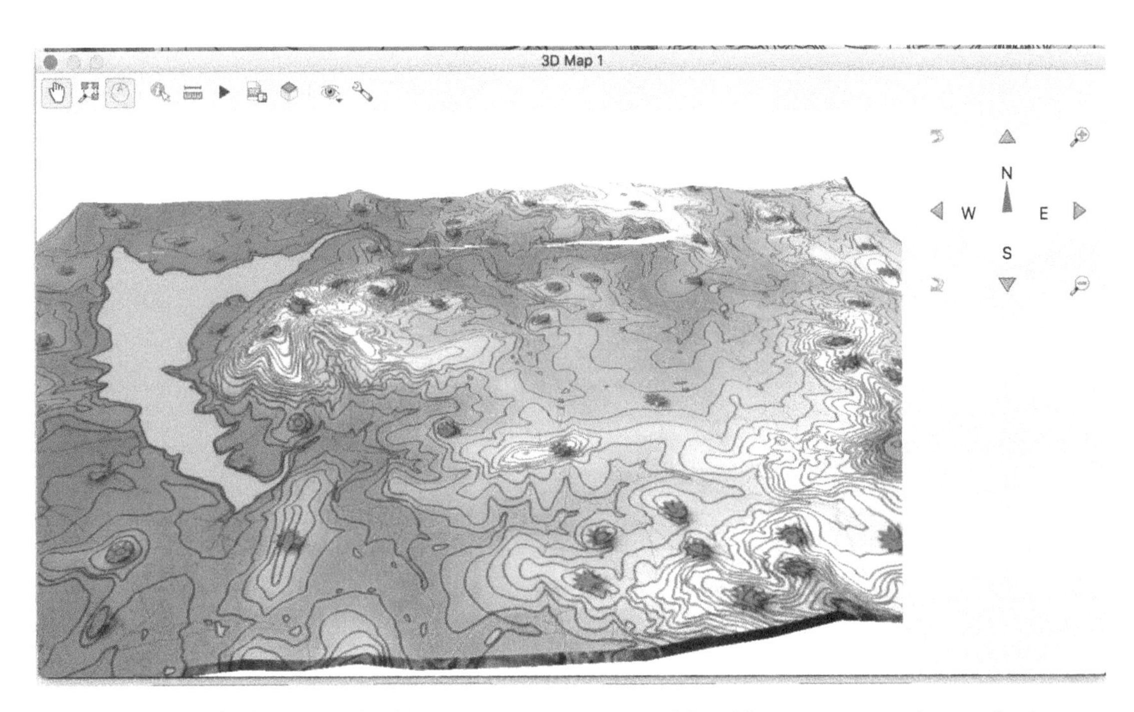

Play around with the controls. You can zoom to extent, identify, measure, and save the image, etc. If you turn off your DEM in the *Layers* panel, and turn on the satellite image `nasa_jpl_clip`, you will see the satellite image overlaid on the DEM, along with the vector files. Whatever you turn off or on in your *Layers* panel, will display in 3D as well.

10.3 Creating and Saving 3D Movies

You can create movies and fly-throughs, by clicking on the *Animations* icon at the top of the 3D window. Then you use the icons shown above the time slider to add points of view, etc. You pick a starting point, save a view, advance and maybe turn to the right, save a view, etc. Try it out.

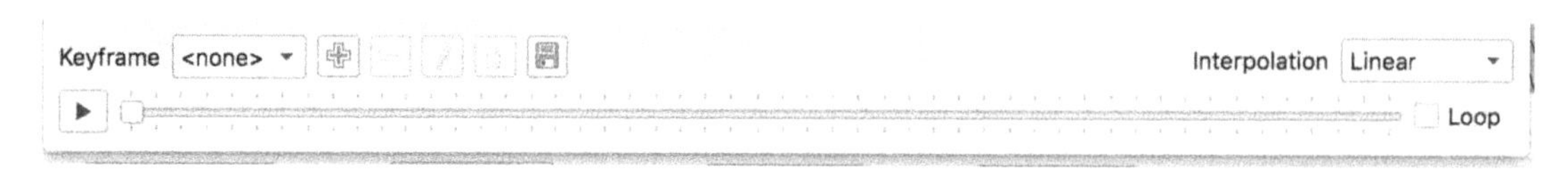

To save your fly-through choose a data type, as shown below:

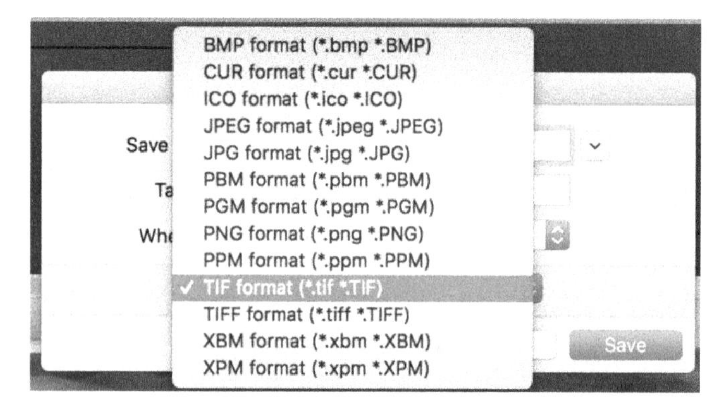

This is very similar to the 3D viewer available in the GRASS GIS, called NVIZ. It is a new capability in QGIS, and will continue to be developed.

Don't forget that you can use this 3D capability to display more than elevation. It is also very useful for displaying data such as population, crime, income, and other demographic information. Simply use a polygon file with attributes such as median income, etc. as your Z value, convert this file to a raster, and you can display it as a 3D map so that high crime neighborhoods are shown with a higher elevation.

11. Georeferencing in QGIS

Satellite images, aerial photos, and modern or historical maps all must be properly referenced to a given map projection in order to line up with your vector GIS data, and georeferencing is the process we use to accomplish this important task. In this exercise, we will be using the QGIS Georeferencer with a high-resolution SPOT image `SPOT_Image_2.jpg` of the Commando Drift study area, acquired in 2012. This was taken from Google Earth as an image export. It is located in:

This image is not georeferenced so we will go through the process to align it with the other layers in our Cape dataset. This is often required for remote sensing imagery, but can also be done with aerial photos, modern, or historical maps as well. For more information on georeferencing historic data, see: `http://geo.nls.uk/urbhist/guides_georeferencingqgis.html`

The georeferencing guide at geo.nls.uk is using QGIS 1.5, so some things look different now, but it gives a sense of how to work with historical maps in QGIS.

11.1 The Georeferencer Dialog

From the main menu, click on `Raster->Georeferencer` to open the *Georeferencer* dialog:

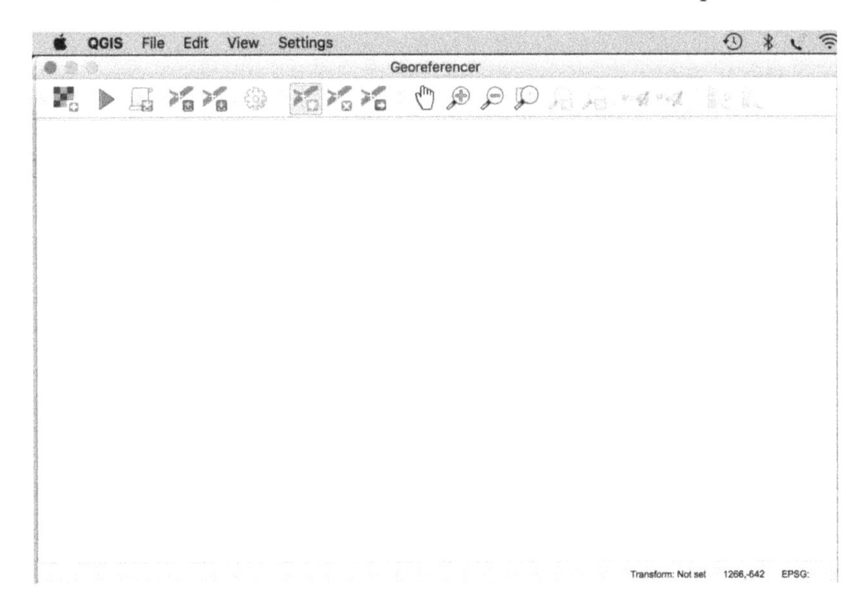

11.2 Georeferencer Icons

You will see the new set of menus and icons along the top that are used to do the steps in the georeferencing process:

Hover your mouse over each to see what it does. Some will be grayed-out as you have no image files open yet. First, add the raster image to be georeferenced by clicking on the *Open Raster* icon and add the `SPOT_Image_2.jpg`, as shown below:

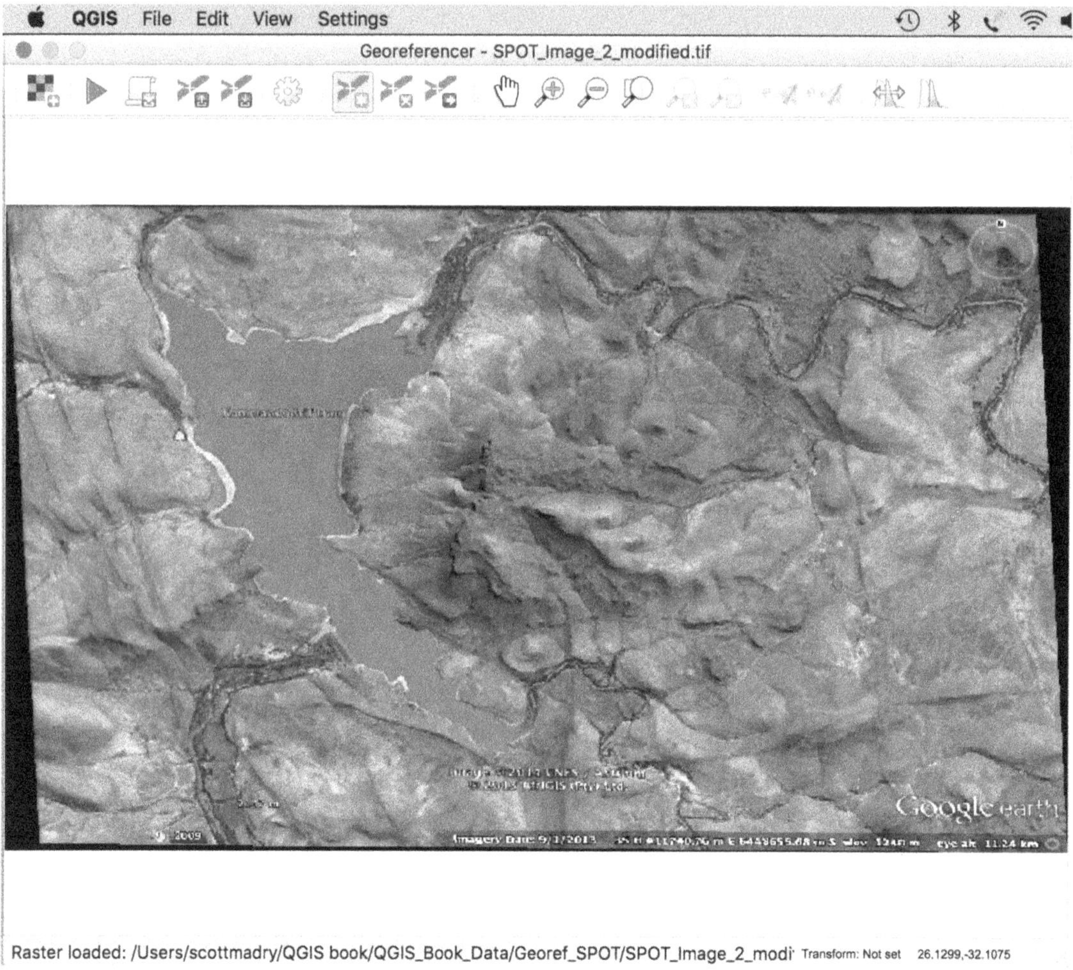

Next, you should click on the *Full Histogram Stretch* to improve the contrast and visibility of the image. As you zoom in around the image, you can click on the *Local Histogram Stretch* icon to do a localized contrast stretch.

You should now have only the `topo_clip.tiff` image visible in the QGIS window (uncheck all the other data) and the `SPOT_Image_2.jpg` in the *Georeferencer* dialog. You may need to right-click on the `topo_clip.tiff` image in the *Layers* panel and reset its transparency (`Properties->Transparen` as you may have made it half transparent in a previous exercise.

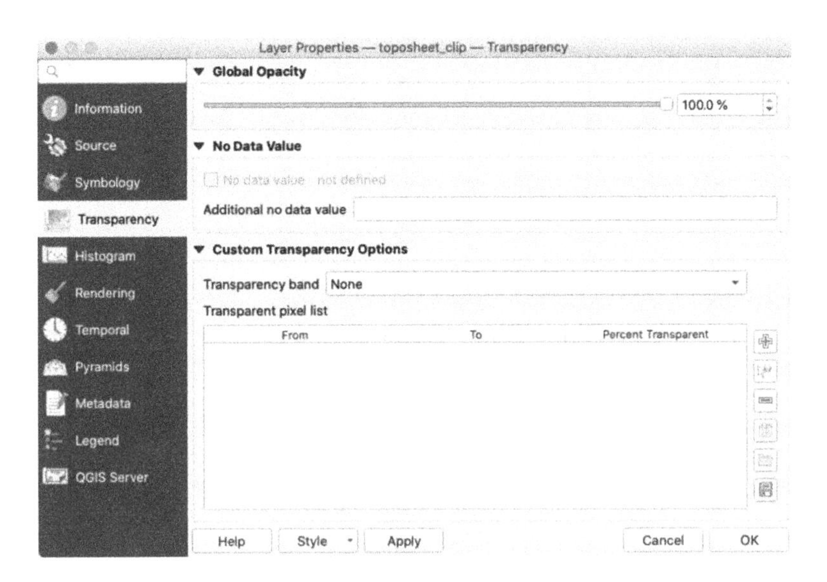

11.3 The Georeferencing Process

Now you will start the georeferencing process. You have the map to be georeferenced and a base map in the map projection and coordinate system that you want it to match. We do this by "point picking", where we identify a location on the SPOT image in the *Georeferencer* dialog, and then find the identical location on the base (`topo_clip.tiff`) image in the QGIS display. We do this several times, all around the images, and then the GIS will mathematically compare the points and warp the SPOT image to overlay correctly on the other image.

Start the process by zooming way in and picking a point in the SPOT image that you can also identify on the other image. Click on one end of the dam of the lake, for example (be careful about using the lake edges, as the height of the water may be different) using the *Add Point* icon.

Once you click to select the point you want to georeference, the *Enter Map Coordinates* dialog will appear where you can either enter the X, Y coordinates or choose directly from the map canvas in the QGIS window.

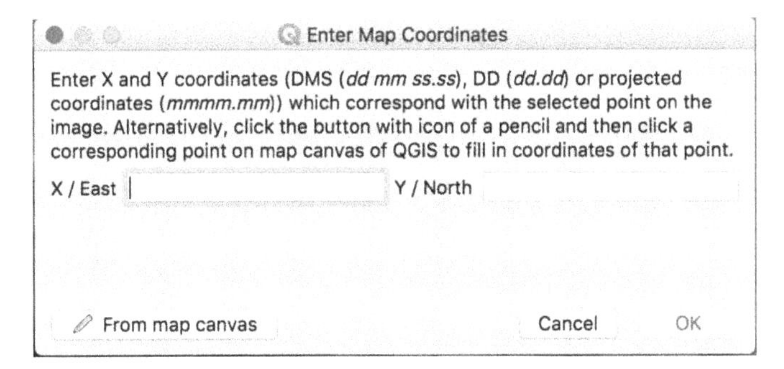

Select *From map canvas* and zoom in to the correct point, then click on *From map canvas* again. The Georeferencer window will reappear again, with the map coordinates entered in the *Enter Map Coordinates* dialog. The coordinates have been filled in:

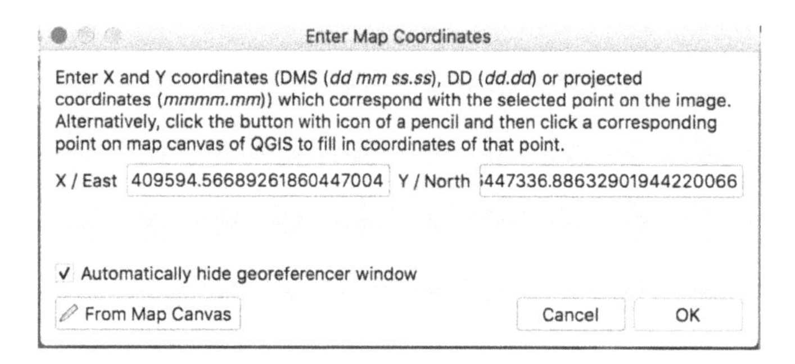

11.4 Point Picking

Click OK at bottom right, and now you will see that a red point has appeared on both images, signifying a point that has been established and georeferenced between the two.

Now click on the *Pan* icon in the *Georeferencer*, and move to another easily identified location, and then move over to the regular QGIS window and zoom in to the same place, and repeat the process.

Once you have at least four points, go to the *Georeferencer* dialog and at top, go to `View->-Panels->GCP table` to make sure it's visible:

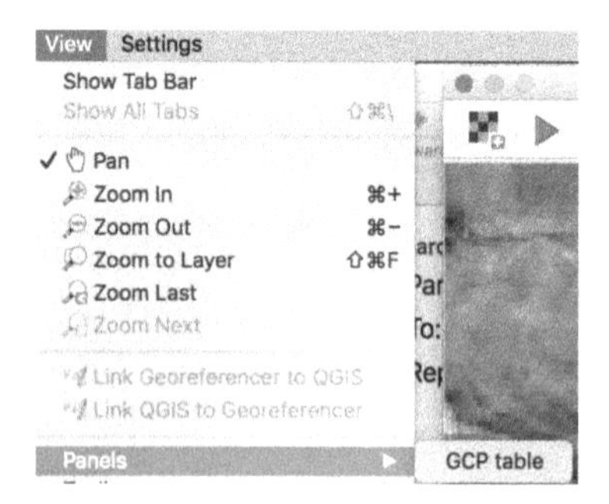

You will see this GCP (Ground Control Point) table at the bottom of the *Georeferencer* dialog:

GCP table

Visible	ID	Source X	Source Y	Dest. X	Dest. Y	dX (pixels)	dY (pixels)	Residual (pixels)
✓	0	26.0415	-32.1066	409595	6.44734e+06	0	0	0
✓	1	26.0483	-32.1052	410205	6.4475e+06	0	0	0
✓	2	26.0282	-32.1167	408313	6.44619e+06	0	0	0
✓	3	26.049	-32.1112	410275	6.44687e+06	0	0	0

Transform: Not set 26.04438,-32.10264

Note that it assigns an ID number, and gives you the X/Y for both the unregistered (*Source*) and registered destination location (*Dest*) location. The residual, or spatial error, is 0.00 because we only have four points so far. Continue to select points between the two images in order to correctly align the SPOT image with the `topo_clip.tiff` image.

11.5 Residual Error and Root Mean Square (RMS) Error

Once you get a minimum of five points, the residual error data are shown for each point, and the total error for all points is shown at bottom right. Here you see the results of the residuals at the bottom of the screen.

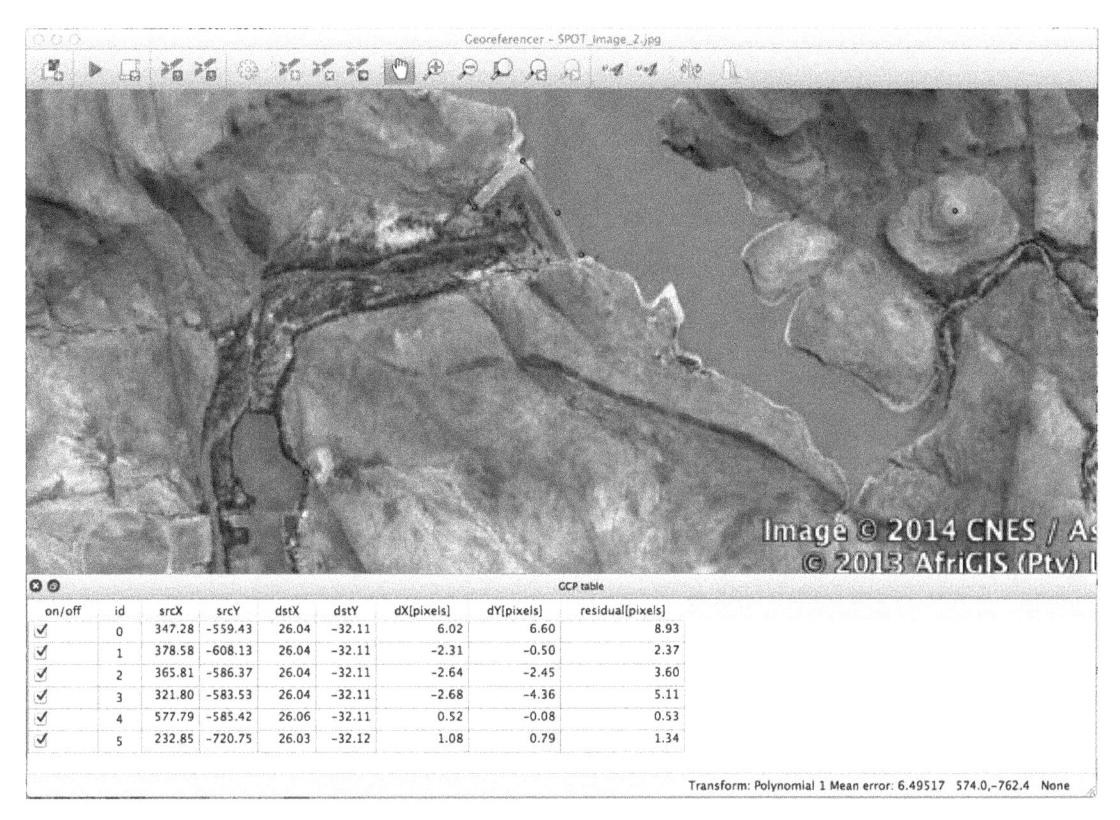

on/off	id	srcX	srcY	dstX	dstY	dX[pixels]	dY[pixels]	residual[pixels]
✓	0	347.28	-559.43	26.04	-32.11	6.02	6.60	8.93
✓	1	378.58	-608.13	26.04	-32.11	-2.31	-0.50	2.37
✓	2	365.81	-586.37	26.04	-32.11	-2.64	-2.45	3.60
✓	3	321.80	-583.53	26.04	-32.11	-2.68	-4.36	5.11
✓	4	577.79	-585.42	26.06	-32.11	0.52	-0.08	0.53
✓	5	232.85	-720.75	26.03	-32.12	1.08	0.79	1.34

You want as small a total RMSE or Root Mean Square Error value as possible. Generally, you always want a value less than the spatial resolution (pixel size) of the image being georeferenced. You can get data about the raster image from the `Settings->Raster Properties` menu in the *Georeferencer* dialog

To compute the RMSE, we compute the residual error for each point, square these values, take the average of the squares, and then take the square root of the average. This is the standard metric for determining how well our georeferencing was done.

You should now pick many more points. You will want a large number of points, with an even distribution of points around the image, but there is no single right number of points or right amount of acceptable error. In general, more points are better, and you want a total RMS error less than the spatial resolution of the image being processed. To learn more about RMS, see: `https://en.wikipedia.org/wiki/Root-mean-square_deviation`

Once you have a large number of points, you can (and should) delete the point(s) with the largest residual numbers, move points, do more points, etc. Iteratively click on the check at far left of each point, and see how the RMS values for the rest change:

GCP table

Visible	ID	Source X	Source Y	Dest. X	Dest. Y	dX (pixels)	dY (pixels)	Residual (pixels)
✓	0	340.606	-439.331	409299	6.44852e+06	-0.589306	-0.570754	0.820391
✓	1	378.529	-609.024	409801	6.44695e+06	-1.94676	-1.88547	2.71014
✓	2	353.627	-567.233	409577	6.44729e+06	3.30935	3.20517	4.60705
✓	3	321.147	-579.142	409224	6.44721e+06	-0.773287	-0.748944	1.07652

Note that you can add points, delete points, and move points using the icons:

You should save and reload your file using these:

11.6 Transformation Settings

Once you have enough points, and a low enough RMS, you should click on the *Transformation Settings* icon, and enter the parameters of the final georeferencing process you wish to use. Additionally, you will enter an *Output raster* name to save your new, georeferenced file for reopening in QGIS. Also, click on *Load in QGIS when done* at bottom left, to see your newly warped image in the regular QGIS display once you run the process.

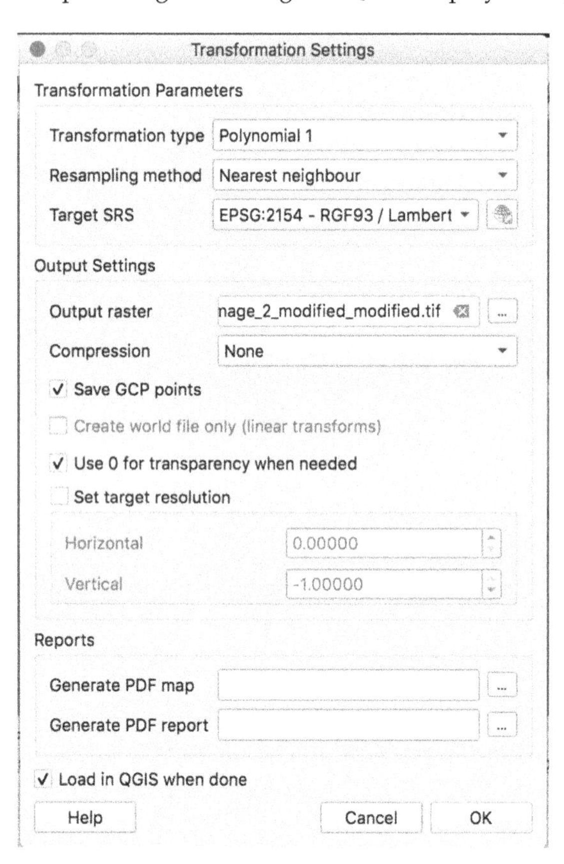

In the *Transformation type* drop-down box, you can select the type of transformation to be used:

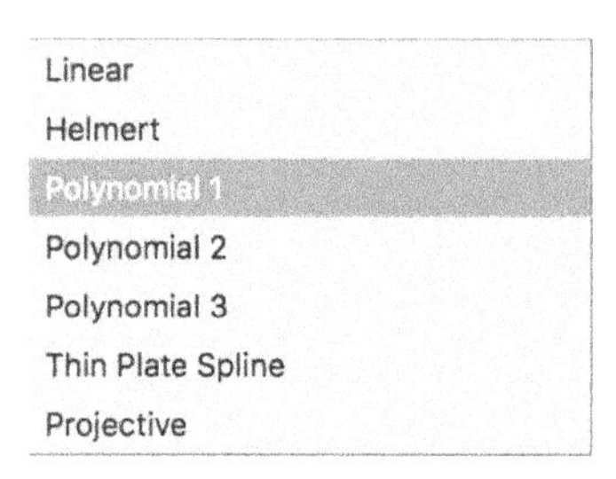

Using different types of transformation are important depending on what type of georeferencing you are attempting, your data, and issues like complex topography. There is extensive literature on how and why to use different transformations. In our case, a simple *Polynomial 1* is fine.

You must also choose a resampling method from these options:

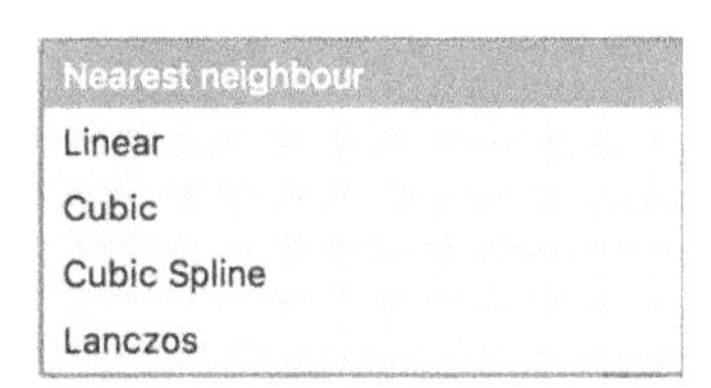

The resampling method also provides different options, but *Nearest Neighbor* is fine. For *Compression*, use *None*, and give the resulting raster file a name. You should always save your points in a file, so you can go back and re-run the analysis if needed.

Click on this icon and fill out the information. As you enter points, you will notice that the residual (pixels) column at the bottom of the screen has a value. This value is the spatial error between the points on the two images. You want the lowest possible number. These numbers will not appear until you fill out the *Transformation Settings* information above, and have at least four control points, as shown below:

GCP table

Visible	ID	Source X	Source Y	Dest. X	Dest. Y	dX (pixels)	dY (pixels)	Residual (pixels)
✓	0	26.0415	-32.1066	409595	6.44734e+06	2.01409	1.22168	2.35564
✓	1	26.0483	-32.1052	410205	6.4475e+06	-1.08876	1.95296	2.23594
✓	2	26.0282	-32.1167	408313	6.44619e+06	-0.500679	-0.00946675	0.500768
✓	3	26.049	-32.1112	410275	6.44687e+06	-0.0702224	-2.08197	2.08316
✓	4	26.0393	-32.096	409397	6.44854e+06	-0.354428	-1.08321	1.13972

Now you can see how well you did, and you can iteratively use the *Delete Points* or *Move GCP Points* icons at top as needed to adjust and improve your RMS. When you have sufficient points and a low enough RMS number, click on the green icon to georeference the image. Be sure you have saved the residuals file, as shown above, so that you can return and continue working on the image if needed. You also need to fully document the process, RMSE, etc. in the metadata for the georeferenced image.

To see the results of your newly georeferenced image, add the new file you just created to the QGIS map window. You need to define a CRS for it. Give it the same as your base map, or in this case, WGS84 UTM35S.

Here is the resulting georeferenced image, made 50% transparent, overlaid on the other image:

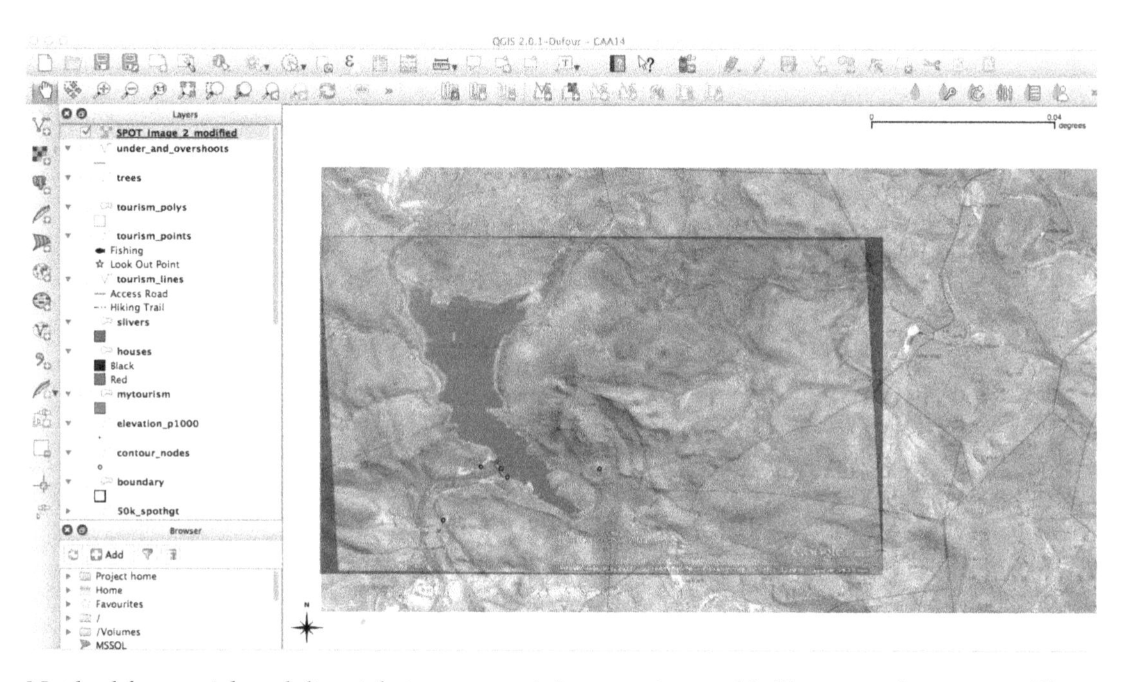

Not bad for a quick and dirty job, just as a training exercise, and it illustrates the process. There is much to learn about proper georeferencing so take the time to learn some before you work with important data. Also keep in mind the concept of error creep, that spatial error here will be propagated to any other data based on it.

The Georeferencer process works well with all kinds of raster data, satellite imagery, aerial photos, historical maps, etc. Aerial photos have a great deal of spatial error, and are more difficult to get precisely lined up, and older maps tend to be less accurate. Once you georeference a map or image you can then vector digitize individual classes of features to populate your GIS with various historical snapshots of your study area. Any feature you can identify on your image or map, such as roads, streams, structures, and vegetation zones can be captured and analyzed.

12. Creating Maps and Reports

Ultimately we need to create properly constructed cartographic maps as a final part of the GIS process. Clear and well-constructed maps are one sign of a professional GIS analyst. We will now make a hardcopy output from our working project. The QGIS *Print Layout* works with the layers that are currently on your map, so you will want to uncheck all but a few vector layers that you want to use.

12.1 The Print Layout Screen

Click on the *New Print Layout* icon on the toolbar or click `command-p` (Mac) or `CTRL-p` (Windows), and it will prompt you to name your new print layout, so you can come back to it when you want:

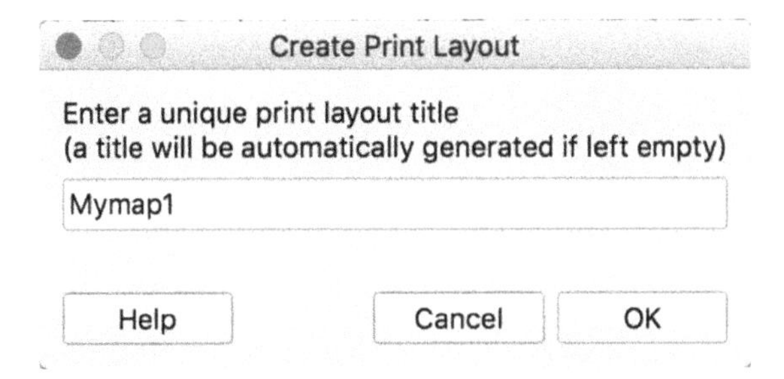

Click *OK* and the map composer window will open:

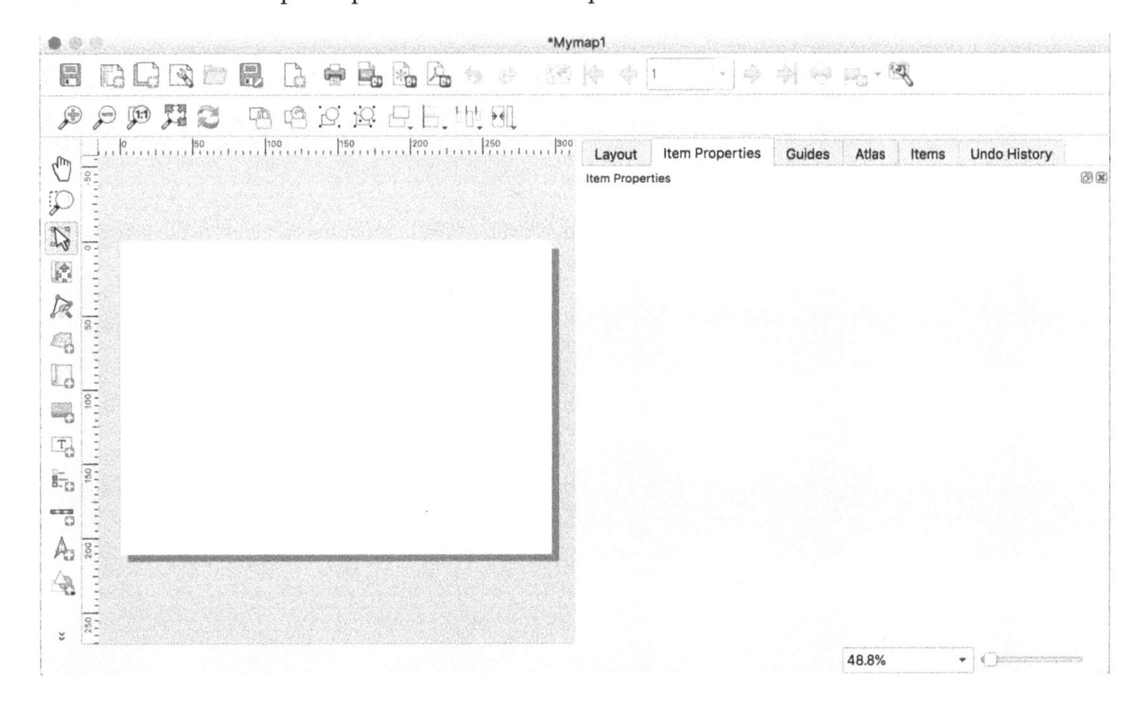

12.2 Print Layout Icons

As with other QGIS windows, the *Print Layout* has its own set of menus and icons at the top.

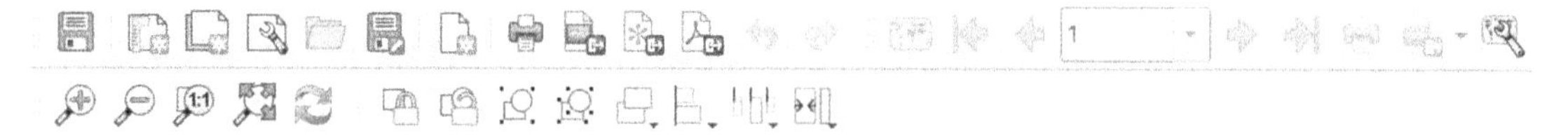

Look at the menu options and hover the cursor over each icon to see what it does. Vertically, along the left side of the *Print Layout* window you'll find the icons used to populate the blank composer page.

12.3 Adding New Maps

The first thing to do is click on the *Add Map* icon on the vertical toolbar. Use your mouse and drag a rectangle on the blank page. Don't cover the entire area, just most of it, leaving a good sized area to top, right, and bottom. You should then see whatever map layers you have open in the QGIS window:

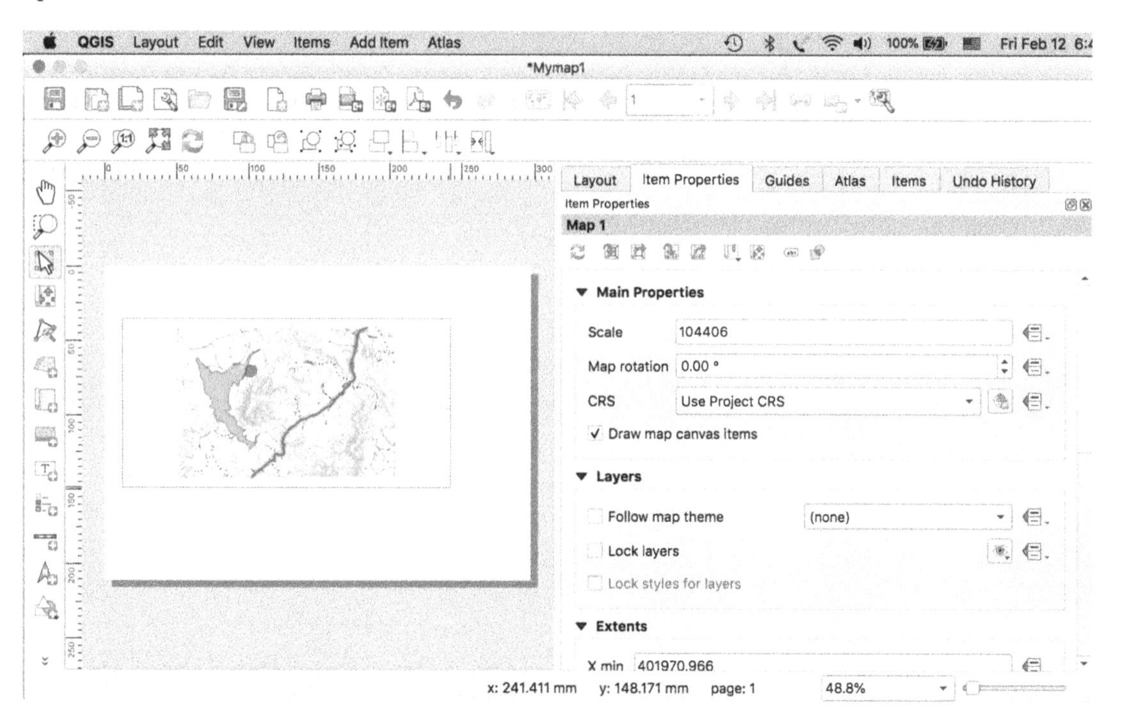

Click on the image and drag the map around, click on a corner and you can alter the size and shape of the map area. Note that you now have a set of options available at right, including the *Item Properties* for the map you just loaded. You can use the border between the map area and the properties tabs to adjust the size by dragging it, thus increasing the size of the map area.

12.4 Changing the Map Composition

To change which layers are displayed, go back to the main QGIS interface, add or delete or add a layer, then come back to the *Print Layout* and click the *Refresh view* button. You will now see the newly updated set of layers.

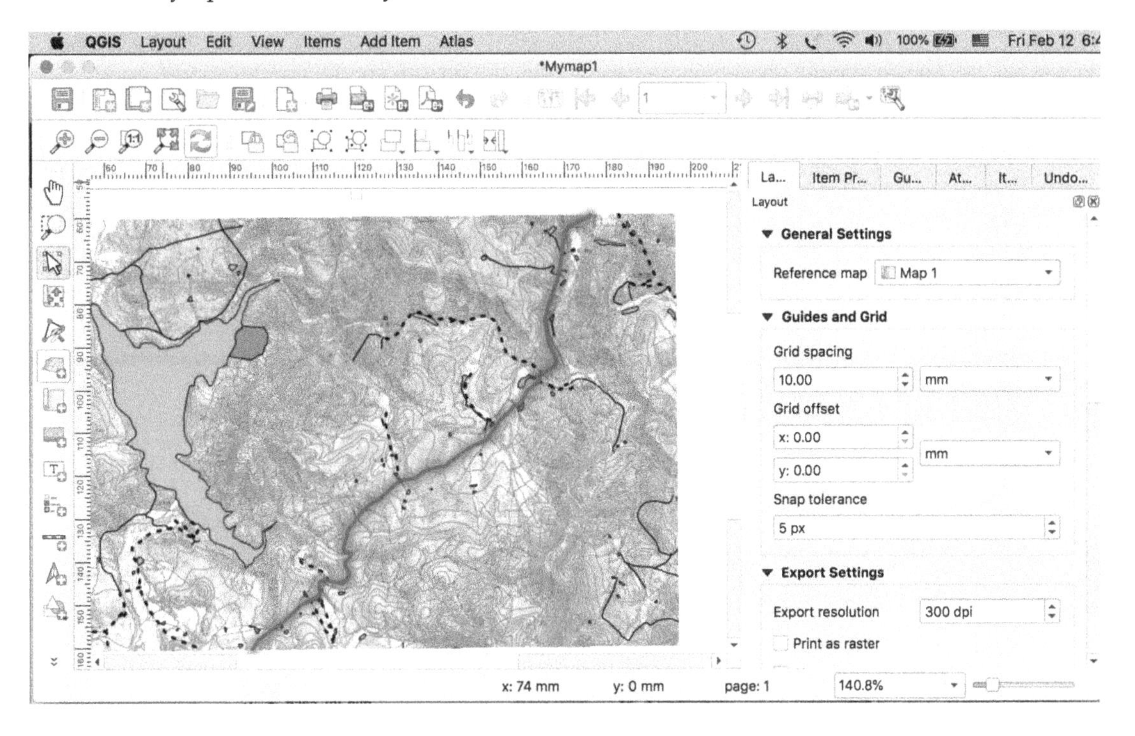

12.5 Map Legends

Click on the *Add Legend* icon at left, and click on the map screen at the top right, and it automatically puts in the vector legend for the files you have displayed.

When you add a legend, all of the layers on your map are included, whether they are visible or not. You will need to remove several of the entries using the *Legend* properties box at right.

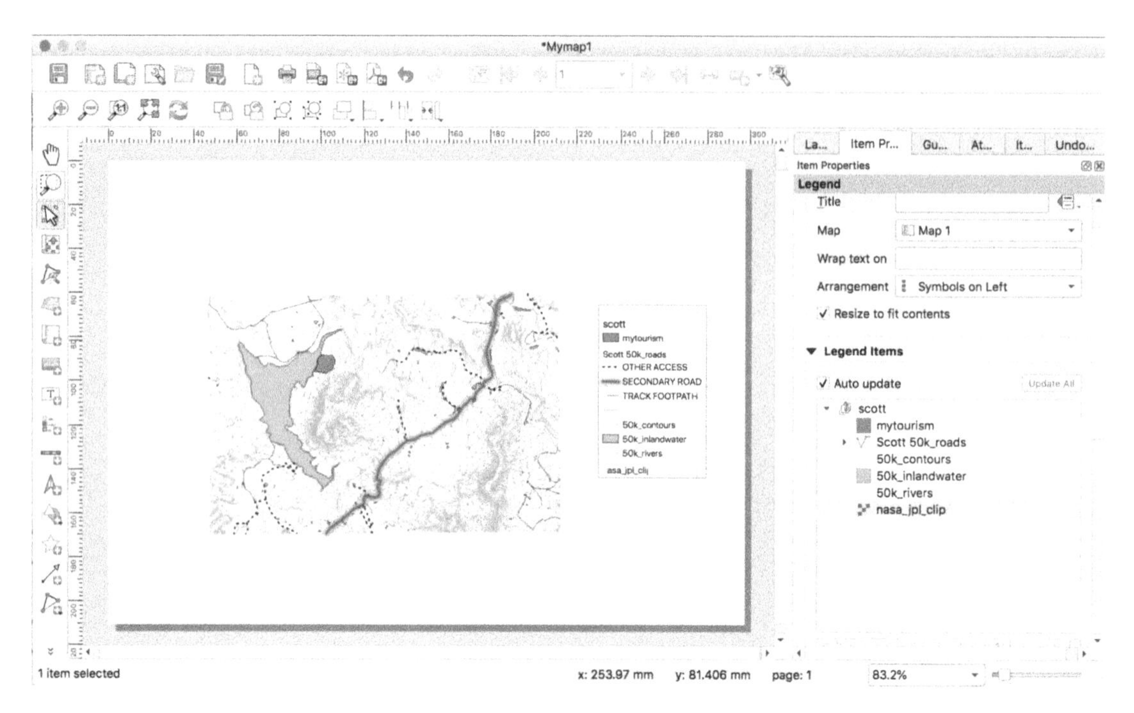

Note the boxes at right change depending on which item you have clicked on in the map space. Click on the legend and the click on *Item Properties* at right, and you will see a set of icons at the bottom of the *Legend Items*:

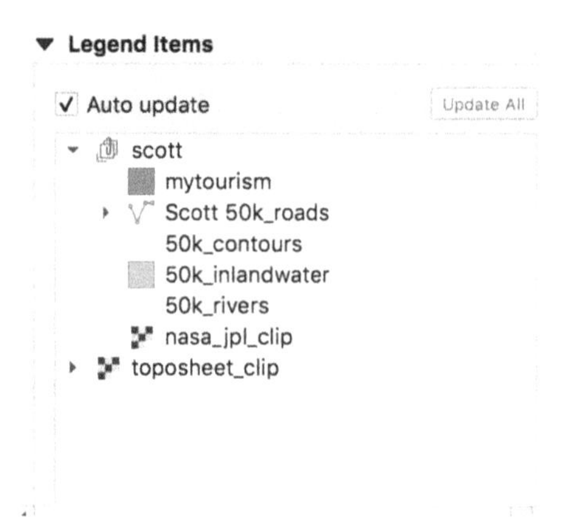

This lets you reorder, add, and delete legends. You may have to resize the map area so everything fits. Again, note that you have different options available at right when you click on the legend. You can delete layers from the legend at the bottom right if there are too many. Uncheck the *Auto update* box at top, click on one of the layers, and click on the red minus icon at bottom to remove it. Do this for several until the legend matches your map at left. You can select several by shift-clicking a group of them. Note that if you click on the map you get the *Map* properties box at right—click on the legend and you get the *Legend* properties.

12.6 Map Labels and Titles

You can add a map label (or title) by clicking on the *Add Label* icon and then clicking or dragging where you want to position the label. If you click, you'll get a dialog box that identifies the location. Click OK, and you can then move it around, resize it, etc. If you drag to create the label, the dialog doesn't—the label is just added to the layout.

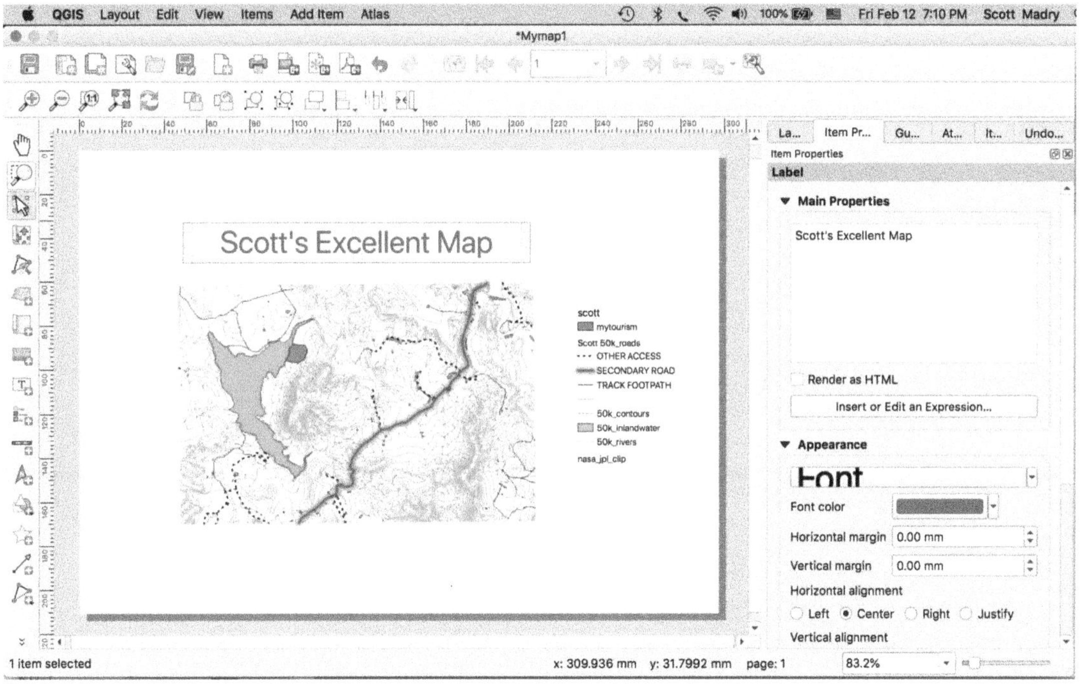

You will want to resize it, so click on *Item Properties* at right, and you will see the options available. Click on *Font*, under *Appearance* to choose the font, style, and size.

When working with your layout, you can always hit `command-z` (Mac) or `CTRL-z` (Windows) to undo one or more steps in the process if you make a mistake.

Every map should have a North arrow and a map scale. A simple way to add a North arrow is to use a capital "N" for the *Title* of your legend, then click on the *Add Arrow* icon  and draw an arrow up through the center of the "N".

There are a couple of other ways to put a North arrow on your map which we'll look at next.

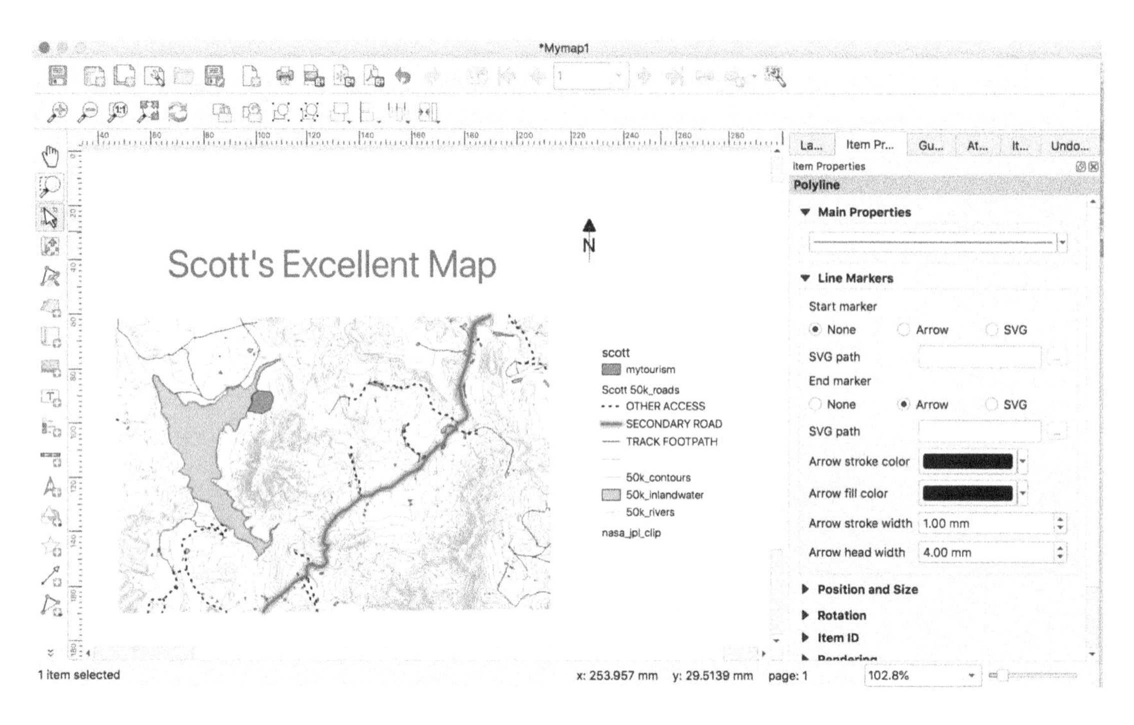

12.7 Adding Images and Photos

We can add both SVG and raster images to our print layout using the *Add Picture* tool . For example, instead of the simple north arrow, you can choose one of several "real" north arrows. *Add Picture* gives you access to a selection of other graphics you can use on your map. Click on the icon and then drag a small square in your layout. Right-click in the new area, and choose *Item Properties*, then choose the arrow style.

A quicker way to add a North arrow is to click on the *Add North Arrow* icon and click on the layout where you want it to appear. You can then adjust the settings using the *Item Properties*.

Add Picture allows us to select a raster image from disk to add to the layout. This can be any valid image format, and once added, you can adjust it using *Item Properties*. Click on *Raster image*, and then the three dots to choose an image from a directory on your computer.

In addition to the SVG icons that come with QGIS, and you can download many other free and open source icons here: `http://wiki.osgeo.org/wiki/OSGeo_map_symbol_set`.

12.8 Scale Bar

Now add the map scale. Click on the *Add Scale Bar* icon and click or drag on the map to automatically place it.

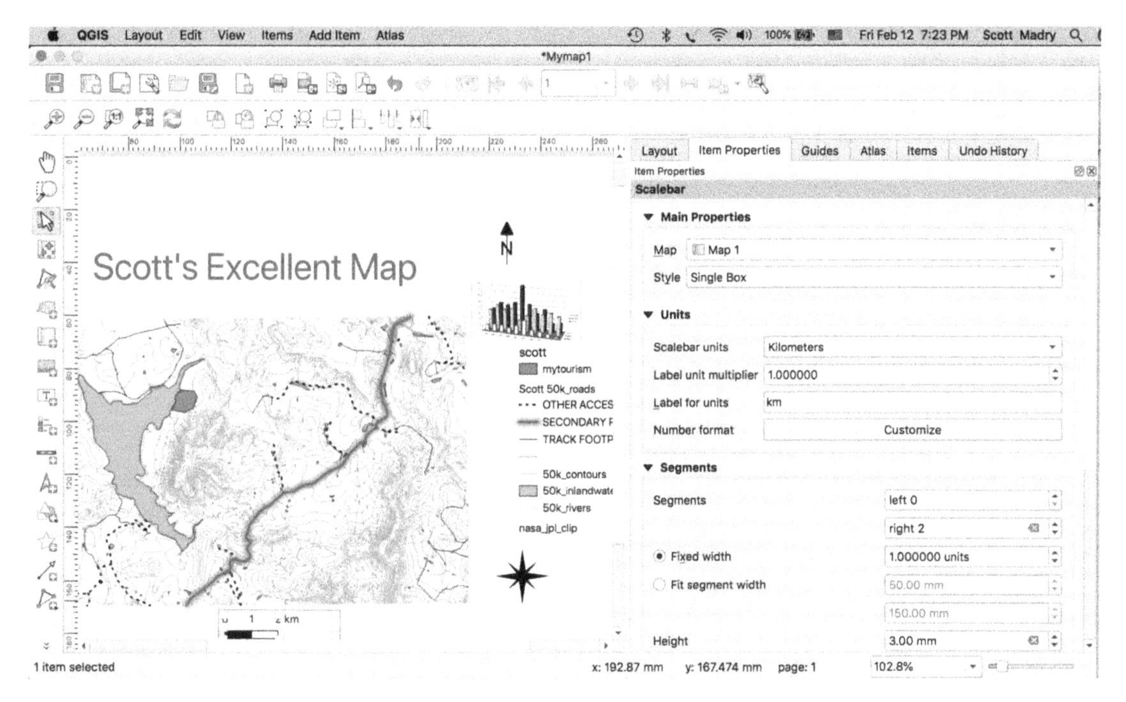

You will see the *Scalebar Item Properties* at right where you can control the style, units, etc. Below you can control the other parameters. This is a lat/long CRS, so pick *Numeric* under the *Style* drop-down box. Click a corner of your map and make it slightly larger and you will see the ratio automatically change. You can move and resize the scalebar to get it just right.

12.9 Coordinate Grids

Now add a coordinate grid. Click on the map image in the print layout, then click on *Item Properties*, and choose *Grids*:

Click on the green plus sign and add a new grid, called *Grid 1*. Click on *Modify Grid* and you will get a new box where you can control the size and style of the grid:

Under *Appearance*, choose an interval of 4000 (meters) X and Y, and you will see your new grid:

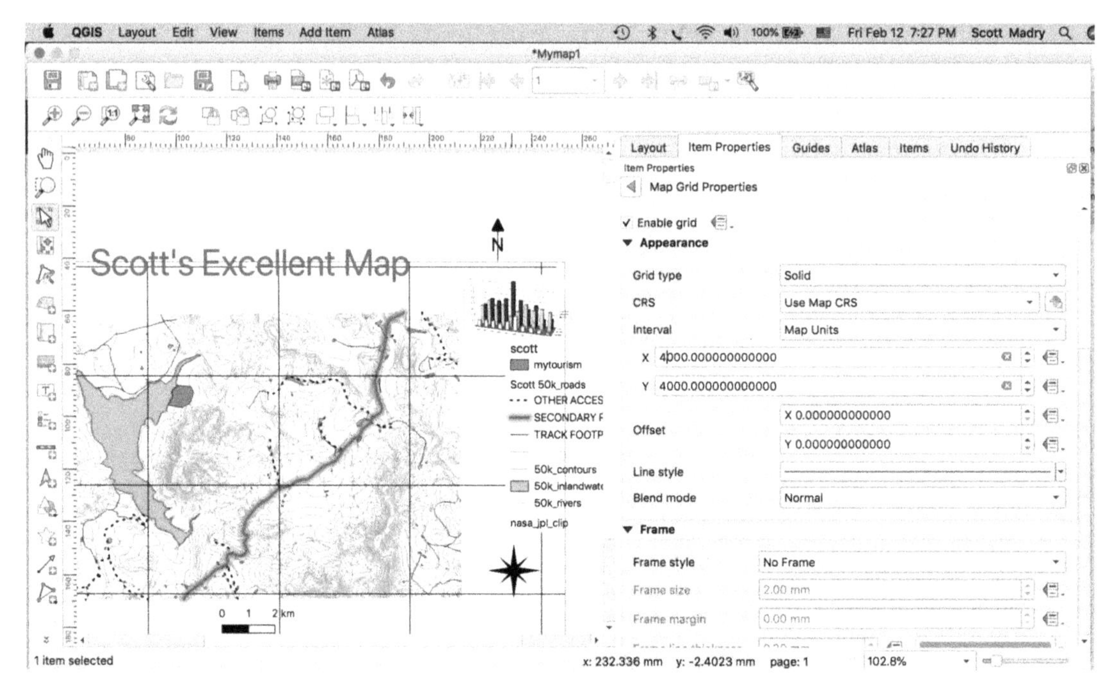

Try changing the *Grid type* to *Cross* and selecting a different *Line style*. Under *FRAME*, choose *Zebra* or one of the other options. Be sure to save your updated map frequently.

12.10 Saving and Printing your Maps

Now you are ready to print your map using the icons along the top of the Layout window. If you need to alter the orientation prior to printing, you can do so by right-clicking on the layout and choosing *Page Properties*.

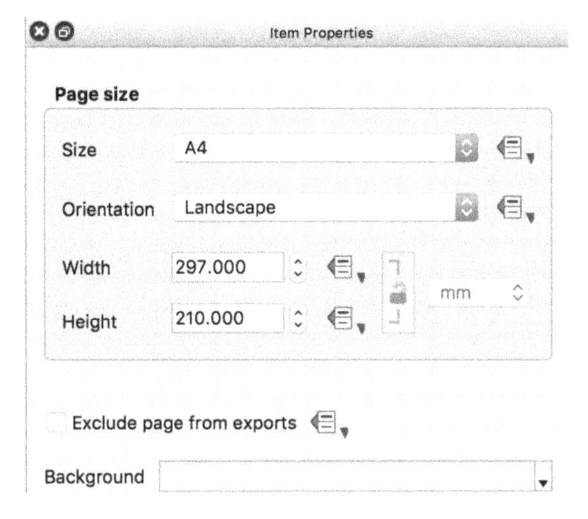

You can print to a printer or export your map layout as JPEG, PDF, or SVG using these icons at the top:

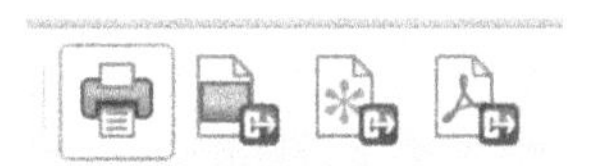

12.11 Map Templates

Once you get a map layout that you like, you can save it as a template using by clicking on . To use an existing template, click on .

There is much, much more to making quality hardcopy maps, but this is enough to get you started. Creating high quality final maps is an art, and is the measure of how your work will often be judged. Create a landscape and portrait map template with your company or university logo and your contact info, then save it for reuse.

12.12 Map Atlas

The QGIS Layout tool has a function to automatically create an atlas map series of your finished cartographic products, as shown below:

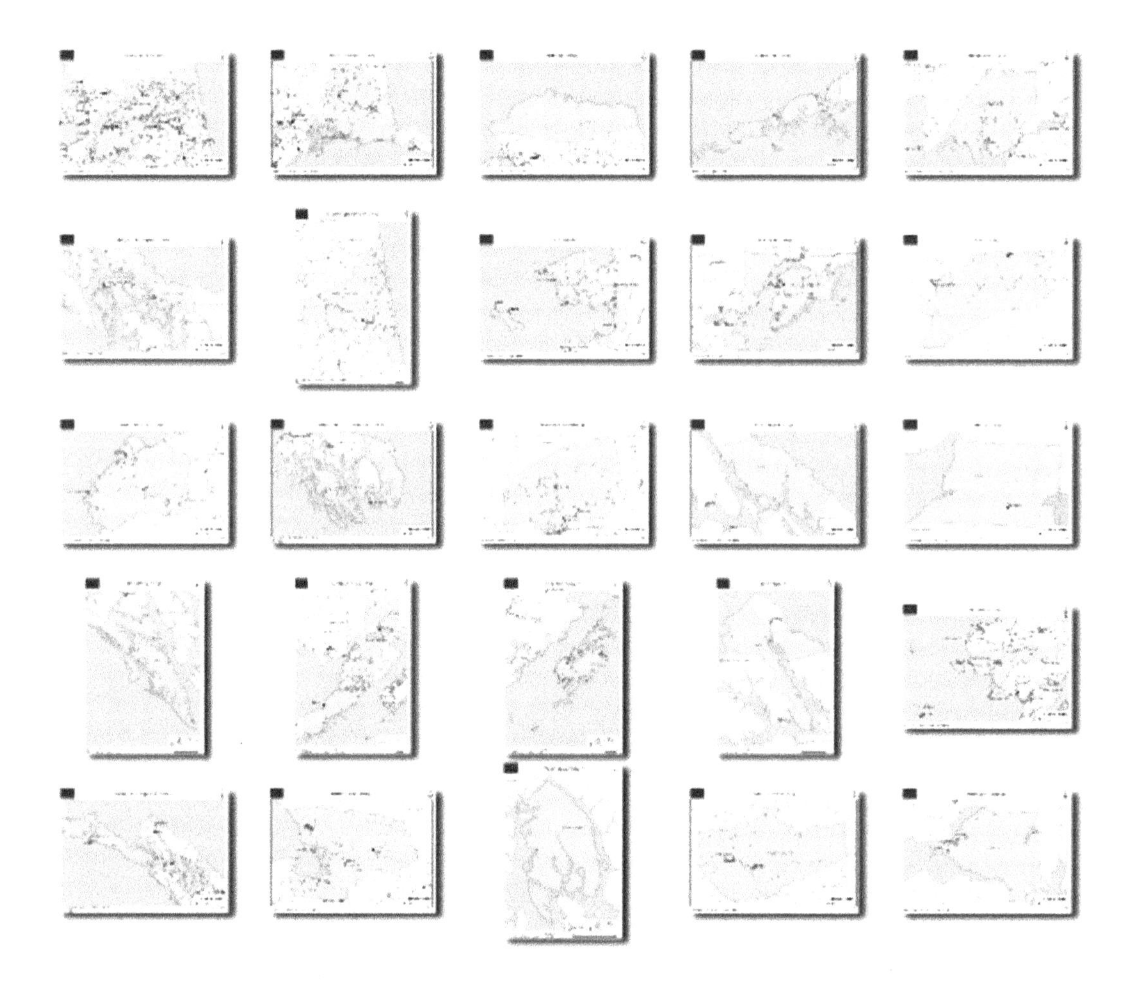

In the *Atlas Settings* icon at the right of the top toolbar, you see the various options for creating and generating an atlas:

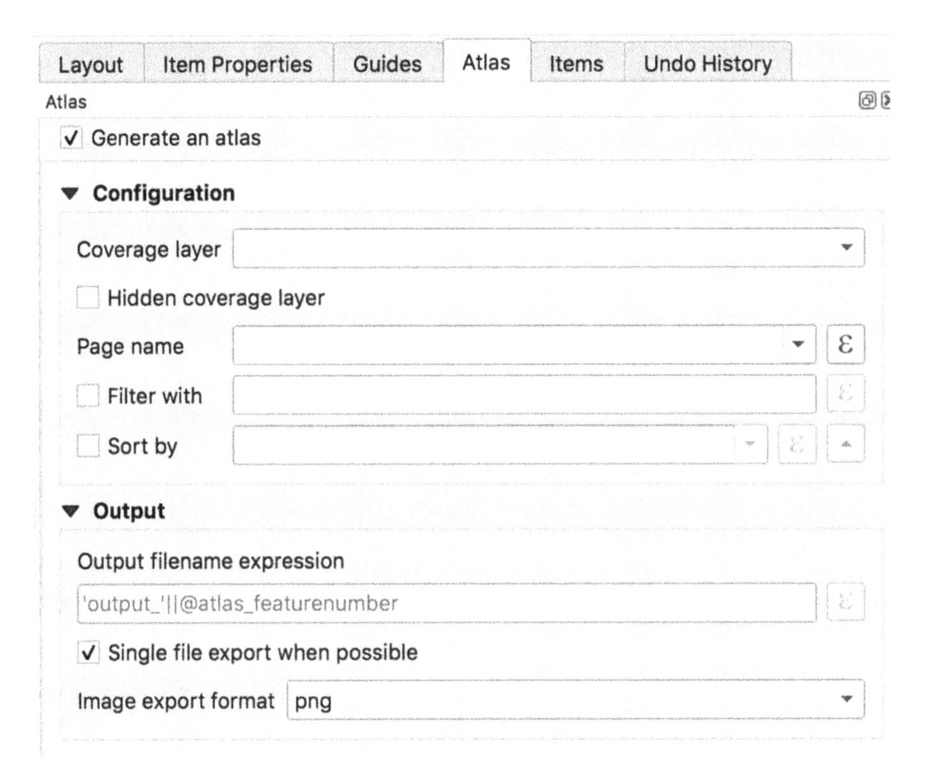

You must choose a vector file that defines the edges and dimensions of the individual tiles to be created, such as the county boundaries of a state. Every page in your atlas will be generated with the same layout template, legend, scale bar, etc. Try it out, and be sure to save your map often.

12.13 QGIS Reports

QGIS 3.0 has a new feature, called QGIS Reports, which lets you set up and format reports and documents that extend the capabilities of QGIS Atlas. From the main QGIS window, click on *Project->New Report* to open the report layout window.

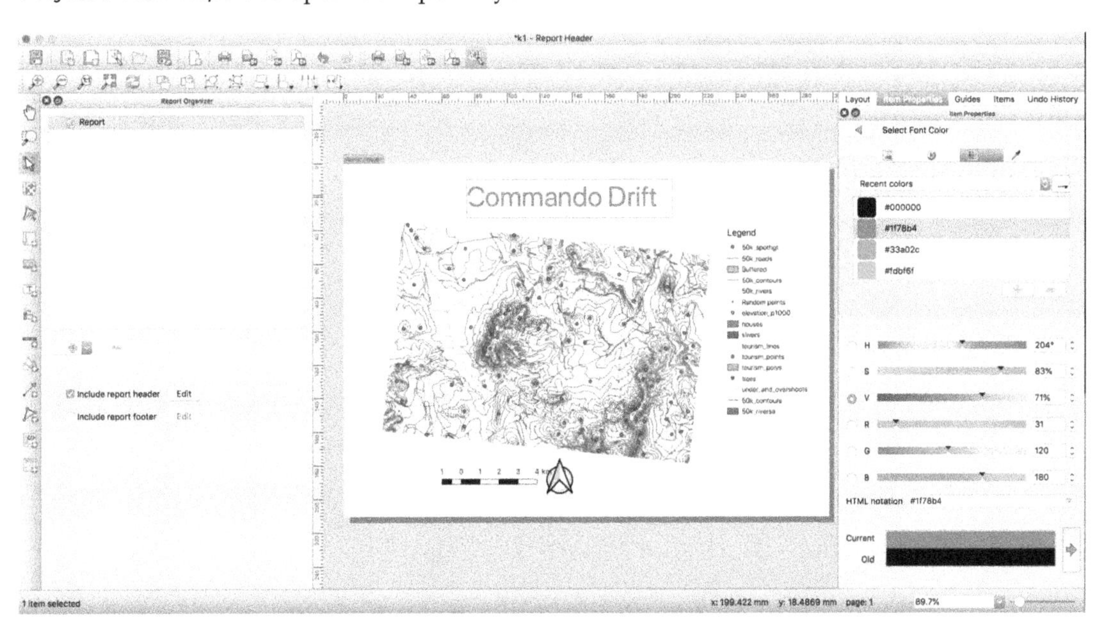

You can create header and footer pages, which can be several pages long. These can be title pages, text, etc. Then you can create static sections and field group sections, each with headers, the body, and footer sections as well. You do this by clicking on the green plus sign below the *Report Organizer*.

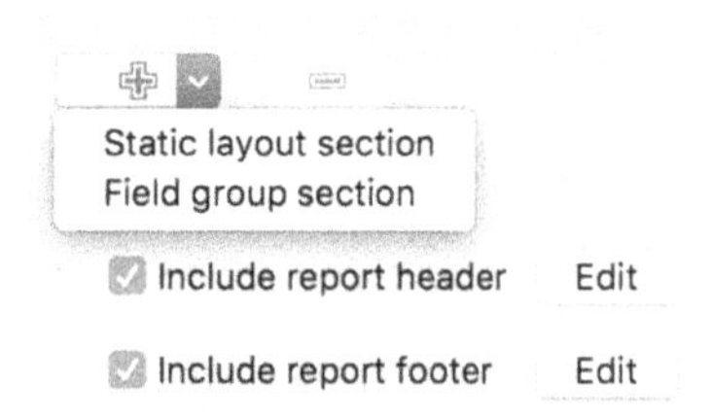

These can include the Atlas function, with a page for each area. The report window at center is initially empty, but by selecting *Report Entry* and *Edit* at left you can add a report header and footer, and then add maps. You set the map up with the icons at left, just as in the *Print Layout*. It will create a report and produce multiple maps like the Atlas function, and add headers and footers. To learn more, see the QGIS Report documentation: `https://docs.qgis.org/3.16/en/docs/user_manual/print_composer/create_reports.html`

Here is a nice video tutorial: `https://www.youtube.com/watch?v=AonZWxDA-cc`

In conclusion, QGIS provides a robust map composition and layout capability, including the production of atlases and embedding graphics directly into reports and documents with QGIS Reports. Learn to use these well, as stated before, your abilities as a GIS analyst will be judged as much by the quality and clarity of your maps as by the analysis that you conduct. Be sure to put your name, North arrow, and scale bar on every map that you produce.

13. GIS and the Chef

Getting started in GIS with QGIS is simple, and there are many different functions that we can use, and many sources of available data. But the end goal is to do some useful analysis or modeling for some practical or research purpose. We ultimately want to ask some relevant questions and provide input into some decision-making process; be that where to route a new highway, redevelop an urban landscape, monitor disease patterns, or model archaeological sites or wildlife habitats in a landscape over time. This is a far more complex process than simply learning a few GIS modules and learning what buttons to click.

Over the years, I have developed a useful analogy to help explain how this process should be approached, both for my students and also for those for whom I am conducting projects.

I would suggest that doing GIS analysis is very much like being a chef or cook in a restaurant. In a restaurant, you have a chef who is responsible for planning the menu and preparing the food for each meal, all within a given budget and timeframe. The chef combines three sets of tools and capabilities to do this. These are: the kitchen, with all its pots and pans and equipment, the raw ingredients and food to be cooked, and finally, the skill and expertise of the chef in turning these into meals ready to be served. Each of these three plays an important role, and, together, they make up each finished plate that is served to the customers.

The chef plans the menu, goes out and purchases the raw vegetables, foods, and spices, does all the prep work, and prepares the meals for that lunch or dinner. Some things can be done early, some are better left to the last minute. The chef uses the tools in the kitchen; the mixers, ovens, knives, and sauté pans, along with his or her knowledge and skills, to turn the raw ingredients in the refrigerator and cupboards into a lovely and satisfying meal for the customers, one that will make them leave happy and want to come back for more.

This is exactly what the trained GIS analyst does. The GIS version of the kitchen, pots, and pans is our QGIS and other, hopefully open source, software. It is our toolkit, with all the various capabilities and functions of our trade. The contents of our refrigerator, the raw food that the chef buys at the market, is our GIS data. We acquire the right data, or create our own, at the right scale and from the right dates, that will let us answer the specific questions at hand.

We then take all of our training and skills, just like the chef, to produce the final products, except that we produce maps, statistics, PowerPoint presentations, reports, or whatever our customer needs, instead of cooking dinner.

What we do is to combine the right tools (QGIS) and the right data, and with the skills of the GIS analyst, produce the finished product, whatever that may be. Someone can come into a restaurant and ask for baked trout, but if the chef did not buy any fish at the market that morning, no matter how much they may want to, the restaurant cannot serve that dish. It is the same with GIS—we must have the right data in order to answer a specific set of questions. If we cannot find the right data, then we have to look at what questions we can address with the data, analysis tools and experience that we have available.

Sometimes the answer is that we just cannot address a given question with what we have, and we have to change our plans or readjust our goals. We just can't cook that dish. Some GIS analysts are Cordon Bleu master chefs, some of us are line cooks with limited training and budgets. Some have more experience, bigger and better larders, and fancier and better equipped kitchens than others. But the process is the same: develop your skills, gather your tools, collect your data, and produce the best products that you possibly can with what you have and what you know.

It is very useful to think about yourself from this perspective when you start a new project, or in creating your career as a GIS professional. You need to, before you begin each project, consider carefully what the goals are, what questions you want to address, what data you will require, what analysis tools you will use, and what resources you will need to produce your finished "meal" on time and on budget. So think about what you need to do to become a well-trained and experienced chef, with the right tools and materials, and get cooking!

14. Useful QGIS Websites

14.1 The QGIS Website and its Importance

As a collaborative, open source project, QGIS is very different from commercial software. First of all, everyone working on the releases and bug fixes are all volunteers, and there are no paid staff in the QGIS organization. So there is no large office, sales and marketing staff, and no toll-free 1-800 telephone number to call for tech support. It is just us, the QGIS community. But this does not mean that there is no help, quite the contrary. In fact, bug fixes and new releases occur on a much faster timeframe than for commercial GIS systems. Everything in this regard is done through the QGIS website at `http://www.qgis.org`.

QGIS

A Free and Open Source Geographic Information System

This is the main portal for all things QGIS. Here you can download the current versions of the code, track updates and changes, reference the users manual, find out about upcoming meetings, get training information, log a bug report, make a donation, share your projects, and more.

It is very important that you become an active member of the community, and if you find a bug, you report it. We in the QGIS community cannot blame a commercial provider if something does not work, it is up to us to fix it, and the first step is to report problems. QGIS actually has a very active bug fixing program, and there are a lot of volunteer help and support channels, chat rooms, etc. We do have an annual bug fix and new development program, where donated funds are used to pay developers to fix known bugs and develop new capabilities. These are bid for, and individuals and groups also use crowdfunding sites to fund the development of needed new tools. We also get private sector funding for specific capabilities.

Listed below are several of the main QGIS websites, blogs, email lists, and more. The important thing is, become a member of the QGIS community, and that start on the QGIS website. It is up to all of us.

As soon as you put together a list of websites things will change, so I am sorry if some links don't work, just do a web search for them.

14.2 QGIS Web Resources

- The main QGIS website: https://www.qgis.org/en/site/
- Downloading the code: https://qgis.org/en/site/forusers/download.html
- The QGIS Support page: https://www.qgis.org/en/site/forusers/support.html with mailing lists, forums, chat, StackExchange, issue tracker, etc.
- The online documentation for version 3.16: https://docs.qgis.org/3.16/en/docs/index.html
- QGIS Plugins: https://plugins.qgis.org/plugins/
- QGIS Case Studies: https://qgis.org/en/site/about/case_studies/index.html including my Burgundy Historical Landscapes project and many others
- QGIS Development page, bugs, issues, and bug reporting: https://www.qgis.org/en/site/getinvolved/development/index#bugs-features-and-issues.html
- The QGIS GitHub: https://github.com/qgis/QGIS
- The QGIS Tutorials and Tips website: http://www.qgistutorials.com/en/index.html
- 3-D Visualization and Analysis of archaeological vector and raster data using open source geospatial software https://library.thehumanjourney.net/659/
- A gentle intro to GIS using QGIS. Developed by Tim Sutton: https://docs.qgis.org/3.16/en/docs/gentle_gis_introduction/index.html
- Intro to QGIS for Archaeological Landscape Analysis by Rebecca Bennett: https://www.academia.edu/2529396/Intro_to_QGIS_for_Archaeological_Landscape_Analysis
- Using the Helmert two-point transformation in QGIS: georeferencing of archaeological site excavation maps in QGIS: https://library.thehumanjourney.net/462/
- Using QGIS and GRASS for processing and analyzing LIDAR data: https://grasswiki.osgeo.org/wiki/Processing_lidar_and_UAV_point_clouds_in_GRASS_GIS_(workshop_at_FOSS4G_Boston_2017)
- QGIS maps and examples of graphics: https://qgis.org/en/site/about/screenshots.html
- Ujaval Gandhi's excellent and comprehensive QGIS Tutorials and Tips: https://www.qgistutorials.com/en/index.html

14.3 QGIS Books, in English, in Print and e-books

There are several excellent QGIS books, many written by the leading minds in the QGIS community: https://www.qgis.org/en/site/forusers/books/index.html

Many of these are published by Locate Press, which is operated by Gary Sherman, the founder of QGIS: https://locatepress.com/books

There are also several QGIS books available in Dutch French, Greek, and Polish as well: `https://www.qgis.org/en/site/forusers/books/index.html`

14.4 QGIS Videos

There are many tutorials for QGIS on YouTube, and more added every week, in a variety of different languages. Simply open YouTube and search for QGIS or QGIS digitizing, etc. For example:

- YouTube videos by Jerrett Totton: `http://www.youtube.com/user/jarretttotton#p/u/4/xcqzEpoRuok`
- QGIS Uncovered by Steven Bernard: `https://www.youtube.com/channel/UCrBM8Ka8HhDAYvQY1VX2P0w/videos`
- MangoMap QGIS Video Tutorials: `http://qgis-tutorials.mangomap.com`

14.5 Online QGIS Curriculum

Top online courses using QGIS: https://www.coursera.org/courses?query=qgis&page=1

14.6 Online GIS Data Archives

- The QGIS Alaska online tutorial datasets: https://qgis.org/downloads/data/
- Geofabrik.de OSM data and shapefiles: https://www.geofabrik.de/data/download.html
- USGS National Map download site: https://www.usgs.gov/core-science-systems/ngp/tnm-delivery/gis-data-download
- Natural Earth global GIS data: http://www.naturalearthdata.com/downloads/
- DIVA national GIS data download: https://www.diva-gis.org/gdata
- ESRI Open Data portal: https://hub.arcgis.com/pages/open-data
- OpenStreetMap data site: https://gisgeography.com/openstreetmap-download-osm-data/
- OpenTopography LiDAR data site: https://opentopography.org
- ESA Sentinal free satellite data: https://scihub.copernicus.eu/dhus
- Natural Earth Global GIS data download (1:10 million): http://www.naturalearthdata.com/downloads/
- David Rumsey historical map collection (historical map downloads): http://www.davidrumsey.com/

14.7 QGIS Blogs:

- Planet QGIS: https://plugins.qgis.org/planet/
- Tim Sutton's new Kartoza blog: http://kartoza.com
- The QGIS blog; http://blog.qgis.org
- Free and Open Source GIS Ramblings by Anita Graser: https://anitagraser.com/tag/qgis/
- The Open Source GIS Blog: http://opensourcegisblog.blogspot.com
- Nyalldawson.net: http://nyalldawson.net

These are only a few of the many web resources that are available to the QGIS and larger open source communities, and they are changing all the time. Feel free to contribute yourself, and if you find an excellent web resource, feel free to contact me and I will add it for the next edition. Happy wandering!

15. Conclusion

QGIS is a very powerful, yet easy to learn and use, comprehensive GIS system that runs on a variety of platforms. It is free, and is open source, and provides a wide range of vector, raster, remote sensing, GPS, database, analysis, and visualization capabilities.

QGIS also allows you to use GRASS, SAGA, GDAL, and Orfeo tools within the QGIS environment, and this makes it an industrial grade powerful system equal to any. There are many more capabilities and functions in QGIS than we have covered here as we have only scratched the surface, but the goal was to get you started enough so that you can continue to learn on your own. See the excellent online QGIS Training Manual (Version 3.4) to continue learning about the many capabilities of QGIS: `https://docs.qgis.org/3.16/en/docs/`

For training materials in other languages, see:

`https://qgis.org/en/site/forusers/trainingmaterial/index.html`

Please remember that all this takes a significant amount of resources, and the QGIS project accepts both annual sponsorships and one-time donations to support the work. If you can, please consider making a contribution, at any level, or participate as a developer or translator if you can. Attend some of the open source GIS conferences, and become an active member of the QGIS community. Most importantly, teach others what you have learned so that more people can use these powerful tools and bring their benefits to our communities and our world. If you have enjoyed this tutorial, or if you feel that you have learned something useful, please show your appreciation with a donation to support the QGIS project. Here is the website:

`https://qgis.org/en/site/getinvolved/donations.html?highlight=donations`

I hope you have found this tutorial helpful, and I wish you good luck in using QGIS for your GIS work. The main thing is to *just do it*. Do some QGIS every day, try something, try something else, make mistakes, try it again, you can't break it, so just do it. Don't forget to back up your files, document what you do, and fill out that metadata!

– Scott Madry, Chapel Hill, NC, USA 2021.

Index

Books from Locate Press

Be sure to visit http://locatepress.com for information on new and upcoming titles.

Discover QGIS 3.x

EXPLORE THE LATEST LONG TERM RELEASE (LTR) OF QGIS!

Discover QGIS 3.x is a comprehensive up-to-date workbook built for both the classroom and professionals looking to build their skills.

Designed to take advantage of the latest QGIS features, this book will guide you in improving your maps and analysis.

Discover QGIS 3.x is an update of the original title, using QGIS 3.6, covering Spatial analysis, Data management, and Cartography. The book includes new exercises and a new section—Advanced Data Visualization.

The book is a complete resource and includes: lab exercises, challenge exercises, all data, discussion questions, and solutions.

Leaflet Cookbook

COOK UP DYNAMIC WEB MAPS USING THE RECIPES IN THE LEAFLET COOKBOOK.

Leaflet Cookbook will guide you in getting started with Leaflet, the leading open-source JavaScript library for creating interactive maps. You'll move swiftly along from the basics to creating interesting and dynamic web maps.

Even if you aren't an HTML/CSS wizard, this book will get you up to speed in creating dynamic and sophisticated web maps. With sample code and complete examples, you'll find it easy to create your own maps in no time.

A download package containing all the code and data used in the book is available so you can follow along as well as use the code as a starting point for your own web maps.

QGIS Map Design - 2nd Edition

LEARN HOW TO USE QGIS 3 TO TAKE YOUR CARTOGRAPHIC PRODUCTS TO THE HIGHEST LEVEL.

QGIS 3.4 opens up exciting new possibilities for creating beautiful and compelling maps!

Building on the first edition, the authors take you step-by-step through the process of using the latest map design tools and techniques in QGIS 3. With numerous new map designs and completely overhauled workflows, this second edition brings you up to speed with current cartographic technology and trends.

See how QGIS continues to surpass the cartographic capabilities of other geoware available today with its data-driven overrides, flexible expression functions, multitudinous color tools, blend modes, and atlasing capabilities. A prior familiarity with basic QGIS capabilities is assumed. All example data and project files are included.

Get ready to launch into the next generation of map design!

The PyQGIS Programmer's Guide

Welcome to the world of PyQGIS, the blending of QGIS and Python to extend and enhance your open source GIS toolbox.

With PyQGIS you can write scripts and plugins to implement new features and perform automated tasks.

This book is updated to work with the next generation of QGIS—version 3.x. After a brief introduction to Python 3, you'll learn how to understand the QGIS Application Programmer Interface (API), write scripts, and build a plugin.

The book is designed to allow you to work through the examples as you go along. At the end of each chapter you will find a set of exercises you can do to enhance your learning experience.

The PyQGIS Programmer's Guide is compatible with the version 3.0 API released with QGIS 3.x and will work for the entire 3.x series of releases.

pgRouting: A Practical Guide

What is pgRouting?

It's a PostgreSQL extension for developing network routing applications and doing graph analysis.

Interested in pgRouting? If so, chances are you already use PostGIS, the spatial extender for the PostgreSQL database management system.

So when you've got PostGIS, why do you need pgRouting? PostGIS is a great tool for molding geometries and doing proximity analysis, however it falls short when your proximity analysis involves constrained paths such as driving along a road or biking along defined paths.

This book will both get you started with pgRouting and guide you into routing, data fixing and costs, as well as using with QGIS and web applications.

Geospatial Power Tools

Everyone loves power tools!

The GDAL and OGR apps are the power tools of the GIS world—best of all, they're free.

The utilities include tools for examining, converting, transforming, building, and analysing data. This book is a collection of the GDAL and OGR documentation, but also includes new content designed to help guide you in using the utilities to solve your current data problems.

Inside you'll find a quick reference for looking up the right syntax and example usage quickly. The book is divided into three parts: *Workflows and examples*, *GDAL raster utilities*, and *OGR vector utilities*.

Once you get a taste of the power the GDAL/OGR suite provides, you'll wonder how you ever got along without them.

See these books and more at http://locatepress.com

CPSIA information can be obtained
at www.ICGtesting.com
Printed in the USA
LVHW070202110122
708281LV00020B/822